Ensemble Algorithms and Their Applications

Ensemble Algorithms and Their Applications

Editors

Panagiotis Pintelas
Ioannis E. Livieris

MDPI • Basel • Beijing • Wuhan • Barcelona • Belgrade • Manchester • Tokyo • Cluj • Tianjin

Editors
Panagiotis Pintelas
University of Patras
Greece

Ioannis E. Livieris
University of Patras
Greece

Editorial Office
MDPI
St. Alban-Anlage 66
4052 Basel, Switzerland

This is a reprint of articles from the Special Issue published online in the open access journal *Algorithms* (ISSN 1999-4893) (available at: https://www.mdpi.com/journal/algorithms/special_issues/Ensemble_Algorithms).

For citation purposes, cite each article independently as indicated on the article page online and as indicated below:

LastName, A.A.; LastName, B.B.; LastName, C.C. Article Title. *Journal Name* **Year**, *Article Number*, Page Range.

ISBN 978-3-03936-958-4 (Hbk)
ISBN 978-3-03936-959-1 (PDF)

© 2020 by the authors. Articles in this book are Open Access and distributed under the Creative Commons Attribution (CC BY) license, which allows users to download, copy and build upon published articles, as long as the author and publisher are properly credited, which ensures maximum dissemination and a wider impact of our publications.
The book as a whole is distributed by MDPI under the terms and conditions of the Creative Commons license CC BY-NC-ND.

Contents

About the Editors .... **vii**

Panagiotis Pintelas and Ioannis E. Livieris
Special Issue on Ensemble Learning and Applications
Reprinted from: *Algorithms* **2020**, *13*, 140, doi:10.3390/a13060140 1

Ioannis E. Livieris, Andreas Kanavos, Vassilis Tampakas, Panagiotis Pintelas
A Weighted Voting Ensemble Self-Labeled Algorithm for the Detection of Lung Abnormalities from X-Rays
Reprinted from: *Algorithms* **2019**, *12*, 64, doi:10.3390/a12030064 5

Konstantinos I. Papageorgiou, Katarzyna Poczeta, Elpiniki Papageorgiou, Vassilis C. Gerogiannis and George Stamoulis
Exploring an Ensemble of Methods that Combines Fuzzy Cognitive Maps and Neural Networks in Solving the Time Series Prediction Problem of Gas Consumption in Greece
Reprinted from: *Algorithms* **2019**, *12*, 235, doi:10.3390/a12110235 21

Emmanuel Pintelas, Ioannis E. Livieris and Panagiotis Pintelas
A Grey-Box Ensemble Model Exploiting Black-Box Accuracy and White-Box Intrinsic Interpretability
Reprinted from: *Algorithms* **2020**, *13*, 17, doi:10.3390/a13010017 49

Stamatis Karlos, Georgios Kostopoulos and Sotiris Kotsiantis
A Soft-Voting Ensemble Based Co-Training Scheme Using Static Selection for Binary Classification Problems
Reprinted from: *Algorithms* **2020**, *13*, 26, doi:10.3390/a13010026 67

Konstantinos Demertzis and Lazaros Iliadis
GeoAI: A Model-Agnostic Meta-Ensemble Zero-Shot Learning Method for Hyperspectral Image Analysis and Classification
Reprinted from: *Algorithms* **2020**, *13*, 61, doi:10.3390/a13030061 87

Kudakwashe Zvarevashe and Oludayo Olugbara
Ensemble Learning of Hybrid Acoustic Features for Speech Emotion Recognition
Reprinted from: *Algorithms* **2020**, *13*, 70, doi:10.3390/a13030070 113

Giannis Haralabopoulos, Ioannis Anagnostopoulos and Derek McAuley
Ensemble Deep Learning for Multilabel Binary Classification of User-Generated Content
Reprinted from: *Algorithms* **2020**, *13*, 83, doi:10.3390/a13040083 137

Ioannis E. Livieris, Emmanuel Pintelas, Stavros Stavroyiannis and Panagiotis Pintelas
Ensemble Deep Learning Models for Forecasting Cryptocurrency Time-Series
Reprinted from: *Algorithms* **2020**, *13*, 121, doi:10.3390/a13050121 151

About the Editors

Panagiotis Pintelas, Ph.D. professor of Computer Science with the Informatics Division of Department of Mathematics at University of Patras, Greece. His research interests include software engineering, AI and ICT in education, machine learning and data mining. He was involved in and directed several dozens of National and European research and development projects. Scholar link: https://scholar.google.gr/citations?user=6UFNG84AAAAJ&hl=el&oi=ao.

Ioannis E. Livieris, Ph.D. holds a Bachelor degree in Mathematics and a Ph.D in Computational Mathematics and Neural Networks from University of Patras. His research includes more than 40 publications in high-level, peer-reviewed journals, 13 volumes in books and 11 peer-reviewed conferences. His scientific interests include neural networks, deep learning, data mining and computational intelligence. Scholar link: https://scholar.google.gr/citations?hl=el&user=0h3_4goAAAAJ.

Editorial

Special Issue on Ensemble Learning and Applications

Panagiotis Pintelas * and Ioannis E. Livieris

Department of Mathematics, University of Patras, 265-00 GR Patras, Greece; livieris@upatras.gr
* Correspondence: ppintelas@gmail.com

Received: 5 June 2020; Accepted: 9 June 2020; Published: 11 June 2020

Abstract: During the last decades, in the area of machine learning and data mining, the development of ensemble methods has gained a significant attention from the scientific community. Machine learning ensemble methods combine multiple learning algorithms to obtain better predictive performance than could be obtained from any of the constituent learning algorithms alone. Combining multiple learning models has been theoretically and experimentally shown to provide significantly better performance than their single base learners. In the literature, ensemble learning algorithms constitute a dominant and state-of-the-art approach for obtaining maximum performance, thus they have been applied in a variety of real-world problems ranging from face and emotion recognition through text classification and medical diagnosis to financial forecasting.

Keywords: ensemble learning; homogeneous and heterogeneous ensembles; fusion strategies; voting schemes; model combination; black, white and gray box models; incremental and evolving learning

1. Introduction

This article is the editorial of the *"Ensemble Learning and Their Applications"* (https://www.mdpi.com/journal/algorithms/special_issues/Ensemble_Algorithms) Special Issue of the *Algorithms* journal. The main aim of this Special Issue is to present the recent advances related to all kinds of ensemble learning algorithms, frameworks, methodologies and investigate the impact of their application in a diversity of real-world problems. The response of the scientific community has been significant, as many original research papers have been submitted for consideration. In total, eight (8) papers were accepted, after going through a careful peer-review process based on quality and novelty criteria. All accepted papers possess significant elements of novelty, cover a diversity of application domains and introduce interesting ensemble-based approaches, which provide readers with a glimpse of the state-of-the-art research in the domain.

During the last decades, the development of ensemble learning methodologies and techniques has gained a significant attention from the scientific and industrial community [1–3]. The basic idea behind these methods is the combination of a set of diverse prediction models for obtaining a composite global model which produces reliable and accurate estimates or predictions. Theoretical and experimental evidence proved that ensemble models provide considerably better prediction performance than single models [4]. Along this line, a variety of ensemble learning algorithms and techniques have been proposed and found their application in various classification and regression real-word problems.

2. Ensemble Learning and Applications

The first paper is entitled *"A Weighted Voting Ensemble Self-Labeled Algorithm for the Detection of Lung Abnormalities from X-Rays"* and it is authored by Livieris et al. [5]. The authors presented a new ensemble-based semi-supervised learning algorithm for the classification of lung abnormalities from chest X-rays. The proposed algorithm exploits a new weighted voting scheme which assigns a vector of weights on each component learner of the ensemble based on its accuracy on each class. The proposed algorithm was extensively evaluated on three famous real-world benchmarks, namely the Pneumonia chest X-rays dataset from Guangzhou Women and Children's Medical Center, the Tuberculosis dataset from Shenzhen Hospital and the cancer CT-medical images dataset. The presented numerical experiments showed the efficiency of the proposed ensemble methodology against simple voting strategy and other traditional semi-supervised methods.

The second paper is authored by Papageorgiou et al. [6] entitled *"Exploring an Ensemble of Methods that Combines Fuzzy Cognitive Maps and Neural Networks in Solving the Time Series Prediction Problem of Gas Consumption in Greece"*. This paper presents an innovative ensemble time-series forecasting model for the prediction of gas consumption demand in Greece. The model is based on an ensemble learning technique which exploits evolutionary Fuzzy Cognitive Maps (FCMs), Artificial Neural Networks (ANNs) and their hybrid structure, named FCM-ANN, for time-series prediction. The prediction performance of the proposed model was compared against that of the Long Short-Term Memory (LSTM) model on three time-series datasets concerning data from distribution points which compose the natural gas grid of a Greek region. The presented results illustrated empirical evidence that the proposed approach could be effectively utilized to forecast gas consumption demand.

The third paper *"A Grey-Box Ensemble Model Exploiting Black-Box Accuracy and White-Box Intrinsic Interpretability"* was written by Pintelas et al. [7]. In this interesting study, the authors proposed a new framework for the development of a Grey-Box machine learning model based on the semi-supervised philosophy. The advantages of the proposed model are that it is nearly as accurate as a Black-Box and it is also interpretable like a White-Box model. More specifically, in their proposed framework, a Black-Box model was utilized for enlarging a small initial labeled dataset, adding the model's most confident predictions of a large unlabeled dataset. In the sequel, the augmented dataset was utilized for training a White-Box model which greatly enhances the interpretability and explainability of the final model (ensemble). For evaluating the flexibility as well as the efficiency of the proposed Grey-Box model, the authors used six benchmarks from three real-world application domains, i.e., finance, education, and medicine. Based on their detailed experimental analysis the authors stated that the proposed model reported comparable and sometimes better prediction accuracy compared to that of a Black-Box while being at the same time interpretable as a White-Box model.

The fourth paper was authored by Karlos et al. [8] entitled *"A Soft-Voting Ensemble Based Co-Training Scheme Using Static Selection for Binary Classification Problems"*. The authors presented an ensemble-based co-training scheme for binary classification problems. The proposed methodology is based on the imposition of an ensemble classifier as a base learner in the co-training framework. Its structure is determined by a static ensemble selection approach from a pool of candidate learners Their experimental results in a variery of classical benchmarks as well as the reported statistical analysis showed the efficacy and efficiency of their approach.

An interesting research entitled *"GeoAI: A Model-Agnostic Meta-Ensemble Zero-Shot Learning Method for Hyperspectral Image Analysis and Classification* was authored by Demertzis and Iliadis [9]. In this work, a new classification model was proposed, named MAME-ZsL (Model-Agnostic Meta-Ensemble Zero-shot Learning), which is based on zero-shot philosophy for geographic object-based scene classification. The attractive advantages of the proposed model are its training stability, its low computational cost, but mostly its remarkable generalization performance thought the reduction of potential overfitting. This is performed by the selection of features which do not cause the gradients to explode or diminish. Additionally, it is worth noticing that the superiority of MAME-ZsL model lies on the fact that the testing set contained instances whose classes were not contained in the training set. The effectiveness of the proposed architecture was presented against state-of-the-art fully supervised deep learning models on two datasets containing images from a reflective optics system imaging spectrometer.

Zvarevashe and Olugbara [10] presented a research paper entitled *"Ensemble Learning of Hybrid Acoustic Features for Speech Emotion Recognition"*. Signal processing and machine learning methods are widely utilized for recognizing human emotions based on extracted features from video files, facial images or speech signals. The authors studied the problem that many classification models were not able to efficiently recognize fear emotion with the same level of accuracy as other emotions. To address this problem, they proposed an elegant methodology for improving the precision of fear and other emotions recognition from speech signals, based on an interesting feature extraction technique. In more detail, their framework extracts highly discriminating speech emotion feature representations from multiple sources which are subsequently agglutinated to form a new set of hybrid acoustic features. The authors conducted a series of experiments on two public databases using a variety of state-of-the-art ensemble classifiers. The presented analysis which reported the efficiency of their approach, provided evidence that the utilization of the new features increased the generalization ability of all ensemble classifiers.

The seventh paper entitled "Ensemble Deep Learning for Multilabel Binary Classification of User-Generated Content is authored by Haralabopoulos et al. [11]. The authors presented an multilabel ensemble model for emotion classification which exploits a new weighted voting strategy based on differential evolution. Additionally, the proposed model used deep learning learners which comprised of convolutional and pooling layers as well as (LSTM) layers which are dedicated for such classification problems. To present the efficiency of their model, they conducted a performance evaluation, on two large and widely used datasets, against state-of-the-art single models and ensemble models which were comprised with the same base learners. The reported numerical experiments showed that the proposed model presented improved classification performance, outperforming state-of-the-art compared models.

Finally, the eighty paper "Ensemble Deep Learning Models for Forecasting Cryptocurrency Time-Series" was authored by Livieris et al. [12]. The main contribution of this research is the combination of three of the most widely employed ensemble strategies: ensemble-averaging, bagging and stacking with advanced deep learning methodologies for forecasting the cryptocurrency hourly prices of Bitcoin, Etherium and Ripple. More analytically, the ensemble models utilized state-of-the-art deep learning models as component learners, which were comprised by combinations of LSTM, Bi-directional LSTM and convolutional layers. The authors conducted an exhaustive experimentation in which the performance of all ensemble deep learning models was compared on both regression and classification problems. The models were evaluated on forecasting of the cryptocurrency price on the next hour (regression) and also on the prediction of next price directional movement (classification) with respect to the current price. Furthermore, the reliability of all ensemble model as well as the efficiency of their predictions was studied by examining for autocorrelation of the errors. The detailed numerical analysis indicated that ensemble learning strategies and deep learning techniques can be efficiently beneficial to each other, and develop accurate and reliable cryptocurrency forecasting models.

We would like to thank the Editor-in-Chief and the editorial office of the *Algorithms* journal for their support and for trusting us with the privilege to edit a special issue in this high-quality journal.

3. Conclusions and Future Approaches

The motivation behind this Special Issue was to make a minor and timely contribution to the existing literature. It is hoped that the novel approaches presented in this Special Issue will be found interesting, constructive and appreciated by the international scientific community. It is also expected that they will inspire further research on innovative ensemble strategies and applications in various multidisciplinary domains. Future approaches may involve exploiting ensemble learning for improving prediction accuracy, machine learning explainability and enhancing model's reliability.

Funding: No funding was provided for this work.

Acknowledgments: The guest editors wish to express their appreciation and deep gratitude to all authors and reviewers which contributed to this Special Issue.

Conflicts of Interest: The guest editors declare no conflict of interest.

References

1. Brown, G. Ensemble Learning. In *Encyclopedia of Machine Learning*; Springer: Boston, MA, USA, 2010; Volume 312.
2. Polikar, R. Ensemble learning. In *Ensemble Machine Learning*; Springer: Boston, MA, USA, 2012; pp. 1–34.
3. Zhang, C.; Ma, Y. *Ensemble Machine Learning: Methods and Applications*; Springer: Boston, MA, USA, 2012.
4. Dietterich, T.G. Ensemble learning. In *The Handbook of Brain Theory and Neural Networks*; MIT Press: Cambridge, MA, USA, 2002; Volume 2, pp. 110–125.
5. Livieris, I.E.; Kanavos, A.; Tampakas, V.; Pintelas, P. A weighted voting ensemble self-labeled algorithm for the detection of lung abnormalities from X-rays. *Algorithms* **2019**, *12*, 64. [CrossRef]
6. Papageorgiou, K.I.; Poczeta, K.; Papageorgiou, E.; Gerogiannis, V.C.; Stamoulis, G. Exploring an Ensemble of Methods that Combines Fuzzy Cognitive Maps and Neural Networks in Solving the Time Series Prediction Problem of Gas Consumption in Greece. *Algorithms* **2019**, *12*, 235. [CrossRef]
7. Pintelas, E.; Livieris, I.E.; Pintelas, P. A Grey-Box Ensemble Model Exploiting Black-Box Accuracy and White-Box Intrinsic Interpretability. *Algorithms* **2020**, *13*, 17. [CrossRef]
8. Karlos, S.; Kostopoulos, G.; Kotsiantis, S. A Soft-Voting Ensemble Based Co-Training Scheme Using Static Selection for Binary Classification Problems. *Algorithms* **2020**, *13*, 26. [CrossRef]
9. Demertzis, K.; Iliadis, L. GeoAI: A Model-Agnostic Meta-Ensemble Zero-Shot Learning Method for Hyperspectral Image Analysis and Classification. *Algorithms* **2020**, *13*, 61. [CrossRef]
10. Zvarevashe, K.; Olugbara, O. Ensemble Learning of Hybrid Acoustic Features for Speech Emotion Recognition. *Algorithms* **2020**, *13*, 70. [CrossRef]
11. Haralabopoulos, G.; Anagnostopoulos, I.; McAuley, D. Ensemble Deep Learning for Multilabel Binary Classification of User-Generated Content. *Algorithms* **2020**, *13*, 83. [CrossRef]
12. Livieris, I.E.; Pintelas, E.; Stavroyiannis, S.; Pintelas, P. Ensemble Deep Learning Models for Forecasting Cryptocurrency Time-Series. *Algorithms* **2020**, *13*, 121. [CrossRef]

© 2020 by the authors. Licensee MDPI, Basel, Switzerland. This article is an open access article distributed under the terms and conditions of the Creative Commons Attribution (CC BY) license (http://creativecommons.org/licenses/by/4.0/).

Article

A Weighted Voting Ensemble Self-Labeled Algorithm for the Detection of Lung Abnormalities from X-Rays

Ioannis E. Livieris [1,*], Andreas Kanavos [1], Vassilis Tampakas [1] and Panagiotis Pintelas [2]

[1] Department of Computer & Informatics Engineering (DISK Lab), Technological Educational Institute of Western Greece, GR 263-34 Antirion, Greece; kanavos@ceid.upatras.gr (A.K.); vtampakas@teimes.gr (V.T.)
[2] Department of Mathematics, University of Patras, GR 265-00 Patras, Greece; ppintelas@gmail.com
* Correspondence: livieris@teiwest.gr

Received: 12 February 2019; Accepted: 13 March 2019; Published: 16 March 2019

Abstract: During the last decades, intensive efforts have been devoted to the extraction of useful knowledge from large volumes of medical data employing advanced machine learning and data mining techniques. Advances in digital chest radiography have enabled research and medical centers to accumulate large repositories of classified (labeled) images and mostly of unclassified (unlabeled) images from human experts. Machine learning methods such as semi-supervised learning algorithms have been proposed as a new direction to address the problem of shortage of available labeled data, by exploiting the explicit classification information of labeled data with the information hidden in the unlabeled data. In the present work, we propose a new ensemble semi-supervised learning algorithm for the classification of lung abnormalities from chest X-rays based on a new weighted voting scheme. The proposed algorithm assigns a vector of weights on each component classifier of the ensemble based on its accuracy on each class. Our numerical experiments illustrate the efficiency of the proposed ensemble methodology against other state-of-the-art classification methods.

Keywords: machine learning; semi-supervised learning; self-labeled algorithms; classifiers; ensemble learning; weighted voting; image classification; lung abnormalities

1. Introduction

The automatic detection of abnormalities, diseases and pathologies constitutes a significant factor in computer-aided medical diagnosis and a vital component in radiologic image analysis. For over a century, radiology has been a typical method for abnormality detection. A typical radiological examination is performed by utilizing a posterior–anterior chest radiograph, which is most commonly called Chest X-Ray (CXR). CXR imaging is widely used for health diagnosis and monitoring, due to its relatively low cost and easy accessibility; thus, it has been established as the single most acquired medical image modality [1]. It constitutes a significant factor for the detection and diagnosis of several pulmonary diseases, such as tuberculosis, lung cancer, pulmonary embolism and interstitial lung disease [1]. However, due to increasing workload pressures, many radiologists today have to daily examine an enormous number of CXRs. Thus, a prediction system trained to predict the risk of specific abnormalities given a particular CXR image is considered essential for providing high quality medical assistance. More specifically, such a decision support system has the potential to support the reading workflow, improve efficiency and reduce prediction errors. Moreover, it could be used to enhance the confidence of the radiologist or prioritize the reading list where critical cases would be read first.

The significant advances in digital chest radiography and the continuously enlarged storage capabilities of electronic media have enabled research centers to accumulate large repositories of classified (labeled) images and mostly of unclassified (unlabeled) images from human experts. To this end, researchers and medical staff were able to leverage and exploit these images by the adoption of machine learning and data mining techniques for the development of intelligent computational systems in order to extract useful and valuable information. As a result, the areas of biomedical research and diagnostic medicine have been dramatically transformed, from rather qualitative sciences which were based on observations of whole organisms to more quantitative sciences which are now based on the extraction of useful knowledge from a large amount of data [2].

Nevertheless, distinguishing the various chest abnormalities from CXRs is a rather challenging task, not only for a prediction model but even for an human expert. The progress in the field has been hampered by the lack of available labeled images for efficiently training a powerful and accurate supervised classification model. Moreover, the process of correctly labeling new unlabeled CXRs usually incurs monetary costs and high time since it constitutes a long and complicated process and requires the efforts of specialized personnel and expert physicians.

Semi-Supervised Learning (SSL) algorithms have been proposed as a new direction to address the problem of shortage of available labeled data, comprising characteristics of both supervised and unsupervised learning algorithms. These algorithms efficiently develop powerful classifiers by meaningfully relating the explicit classification information of labeled data with the information hidden in the unlabeled data [3,4]. Self-labeled algorithms probably constitute the most popular class of SSL algorithms due to their simplicity of implementation, their wrapper-based philosophy and good classification performance [2,5–8]. This class of algorithms exploits a large amount of unlabeled data via a self-learning process based on supervised learners. In other words, they perform an iterative procedure, enriching the initial labeled data, based on the assumption that their own predictions tend to be correct.

Recently, Triguero et al. [9] proposed an in-depth taxonomy based on the main characteristics presented in them and conducted a comprehensive research of their classification efficacy on several datasets. Generally, self-labeled algorithm can be classified in two main groups: *Self-training* and *Co-training*. In the original Self-training [10], a single classifier is iteratively trained on an enlarged labeled dataset with its most confident predictions on unlabeled data while in Co-training [11], two classifiers are separately trained utilizing two different views on a labeled dataset and then each classifier augments the labeled data of the other with its most confident predictions on unlabeled data. Along this line, several self-labeled algorithms have been proposed in the literature, while some of them exploit ensemble methodologies and techniques.

Democratic-Co learning [12] is based on an ensemble philosophy since it uses three independent classifiers following a majority voting and a confidence measurement strategy for predicting the values of unlabeled examples. Tri-training algorithm [13] utilizes a bagging ensemble of three classifiers which are trained on data subsets generated through bootstrap sampling from the original labeled set and teach each other using on majority voting strategy. Co-Forest [14] utilizes bootstrap sample data from the labeled set in order to train Random trees. At each iteration, each random tree is reconstructed by newly selected unlabeled instances for its concomitant ensemble, utilizing a majority voting technique. Co-Bagging [15] trains multiple base classifiers on bootstrap data created by random resampling with replacement from the training set. Each bootstrap sample contains about 2/3 of the original training set, where each example can appear multiple times. Recently, a new approach has been given by Livieris et al. [2,8,16,17] and Livieris [18] in which some ensemble self-labeled algorithms are proposed based on voting schemes. The proposed algorithms exploit the individual predictions of the most efficient and frequently used self-labeled algorithms using simple voting methodologies.

Motivated by these works, we propose a new semi-supervised self-labeled algorithm which is based on a sophisticated ensemble philosophy. The proposed algorithm exploits the individual predictions of self-labeled algorithms, using a new weighted voting methodology. The proposed weighted strategy assigns weights on each component classifier of the ensemble based on its accuracy on each class. Our main aim is to measure the effectiveness of our weighted voting ensemble scheme over the majority voting ensembles, using identical component classifiers in all cases. On top of that, we want to verify that powerful classification models could be developed by the adaptation of advanced ensemble methodologies in the SSL framework. Our preliminary numerical experiments prove the efficiency and the classification accuracy of the proposed algorithm, demonstrating that reliable prediction models could be developed by incorporating ensemble methodologies in the semi-supervised framework.

The remainder of this paper is organized as follows: Section 2 presents a brief survey of recent studies concerning the application of machine learning for the detection of lung abnormalities from X-rays. Section 3 presents a detailed description of the proposed weighted voting scheme and ensemble algorithm. Section 4 presents a series of experiments carried out in order to examine and evaluate the accuracy of the proposed algorithm against the most popular self-labeled classification algorithms. Finally, Section 5 discusses the conclusions and some research topics for future work.

2. Related Work

The significance of medical imaging for the diagnosis of diseases has been established for the treatment of chest pathologies and their early detection. During the last decades, the advances of digital technology and chest radiography as well as the rapid development of digital image retrieval have renewed the progress in new technologies for the diagnosis of lung abnormalities. More specifically, research has been focused on the development of Computer-Aided Diagnostic (CAD) models for abnormality detection in order to assist medical staff. Along this line, a variety of methodologies have been proposed based on machine learning techniques, aiming on classifying and/or detecting abnormalities in patients' medical images. A number of studies have been carried out in recent years; some useful outcomes of them are briefly presented below.

Jaeger et al. [19] proposed a CAD system for tuberculosis in conventional posteroanterior chest radiographs. Their proposed model initially utilizes a graph cut segmentation method to extract the lung region from the CXRs and then a set of texture and shape features in the lung region is computed in order to classify the patient as normal or abnormal. Their extensive numerical experiments on two real-world datasets illustrated the efficiency of the proposed CAD system for tuberculosis screening, achieving higher performance compared to that of human readings.

Melendez et al. [20] recommend a novel CAD system for detecting tuberculosis on chest X-rays based on multiple-instance learning. Their proposed system is based on the idea of utilizing probability estimations, instead of the sign of a decision function, to guide the multiple-instance learning process. Furthermore, an advantage of their method is that it does not require labeling of each feature sample during the training process but only a global class label characterizing a group of samples.

Alam et al. [21] utilized a multi-class support vector machine classifier and developed an efficient lung cancer detection and prediction model. The image enhancement and image segmentation have been done independently in every stage of the classification process. Image scaling, color space transformation and contrast enhancement have been utilized for image enhancement while threshold and marker-controlled watershed have been utilized for segmentation. In the sequel, the support vector machine classifier categorizes a set of textural features extracted from the separated regions of interest. Based on their numerical experiments, the authors concluded that the proposed algorithm can efficiently detect a cancer-affected cell and its corresponding stage such as initial, middle, or final. Furthermore, in case no cancer-affected cell is found in the input image then it checks the probability of lung cancer.

In more recent works, Madani [22] focused on the detection of abnormalities in chest X-ray images, having available only a fairly small size dataset of annotated images. Their proposed method deals with both problems of labeled data scarcity and data domain overfitting, by utilizing Generative Adversarial Networks (GAN) in a SSL architecture. In general, GAN utilize two networks: a generator which seeks to create as realistic images as possible and a discriminator which seeks to distinguish between real data and generated data. Next, these networks are involved in a minimax game to find the Nash equilibrium between them. Based on their experiments, the author concluded that the annotation effort is reduced considerably to achieve similar performance through supervised training techniques.

In [2], Livieris et al. evaluated the classification efficacy of an ensemble SSL algorithm, called CST-Voting, for CXR classification of tuberculosis. The proposed algorithm combines the individual predictions of three efficient self-labeled algorithms i.e., Co-training, Self-training and Tri-training using a simple majority voting methodology. The authors presented some interesting results, illustrating the efficiency of the proposed algorithm against several classical algorithms. Additionally, their experiments lead them to the conclusion that reliable and robust prediction models could be developed utilizing a few labeled and many unlabeled data. In [16] the authors extended the previous work and proposed DTCo algorithm for the classification of X-rays. The proposed ensemble algorithm exploits the predictions of Democratic-Co learning, Tri-training and Co-Bagging utilizing a maximum-probability voting scheme. Along this line, Livieris et al. [17] proposed EnSL algorithm which constitutes a generalized scheme of the previous works. More specifically, EnSL constitutes a majority voting scheme of N self-labeled algorithms. Their preliminary numerical experiments demonstrated that robust classification models could be developed by the adaptation of ensemble methodologies in the SSL framework.

Guan and Huang [23] considered the problem of multi-label thorax disease classification on chest X-ray images by proposing a Category-wise Residual Attention Learning (CRAL) framework. CRAL predicts the presence of multiple pathologies in a class-specific attentive view, aiming to suppress the obstacles of irrelevant classes by endowing small weights to the corresponding feature representation while the same time, the relevant features would be strengthened by assigning larger weights. More analytically, their proposed framework consists of two modules: feature embedding module and attention learning module. The feature embedding module learns high-level features using a neural network classifier while the attention learning module focuses on exploring the assignment scheme of different categories. Based on their numerical experiments, the authors stated that their proposed methodology constitutes a new state of the art.

3. A New Weighted Voting Ensemble Self-Labeled Algorithm

In this section, we present a detailed description of the proposed self-labeled algorithm, which is based on an ensemble philosophy, entitled Weighed voting Ensemble Self-Labeled (WvEnSL) algorithm.

Generally, the generation of an ensemble of classifiers considers mainly two steps: *Selection* and *Combination*. The selection of the component classifiers is considered essential for the efficiency of the ensemble and the key point for its efficacy is based on their diversity and their accuracy; while the combination of the individual classifiers' predictions takes place through several techniques with different philosophy [24,25].

By taking these into consideration, the proposed algorithm is based on the idea of selecting a set $C = (C_1, C_2, \ldots, C_N)$ of N self-labeled classifiers by applying different algorithms (with heterogeneous model representations) to a single dataset and the combination of their individual predictions takes place through a new weighted voting methodology. It is worth noticing that weighted voting is a commonly used strategy for combining predictions in pairwise classification in which the classifiers are not treated equally. Each classifier is evaluated on a evaluation set D and associated with a coefficient (weight), usually proportional to its classification accuracy.

Let us consider a dataset D with M classes, which is utilized for the evaluation of each component classifier. More specifically, the performance of each classifier C_i, with $i = 1, 2, \ldots, N$ is evaluated on D and a $N \times M$ matrix W is defined, as follows

$$W = \begin{bmatrix} w_{1,1} & w_{1,2} & \cdots & w_{1,M} \\ w_{2,1} & w_{2,2} & \cdots & w_{2,M} \\ \vdots & \vdots & \ddots & \vdots \\ w_{N,1} & w_{N,2} & \cdots & w_{N,M} \end{bmatrix}$$

where each element $w_{i,j}$ is defined by

$$w_{i,j} = \frac{2p_j^{(C_i)}}{|D_j| + p_j^{(C_i)} + q_j^{(C_i)}}, \tag{1}$$

where D_j is the set of instances of the dataset belonging to the class j, $p_j^{(C_i)}$ are the number of correct predictions of classifier C_i on D_j and $q_j^{(C_i)}$ are the number of incorrect predictions of C_i that an instance belongs to class j. Clearly, each weight $w_{i,j}$ is the F_1-score of classifier C_i for j class [26]. The rationale behind (1) is to measure the efficiency of each classifier, relative to each class j of the evaluation set D.

Subsequently, the class $\hat{y}$ of each unknown instance x in the test set is computed by

$$\hat{y} = \arg\max_j \sum_{i=1}^{N} w_{i,j} \chi_A(C_i(x) = j),$$

where function $\arg\max$ returns the value of index corresponding to the largest value from array, $A = \{1, 2, \ldots, M\}$ is the set of unique class labels and χ_A is the characteristic function which takes into account the prediction $j \in A$ of a classifier C_i on an instance x and creates a vector in which the j coordinate takes a value of one and the rest take the value of zero. At this point, it is worth mentioning that in our implementation we selected to evaluate the performance of each classifier of the ensemble on the initial training labeled set L.

A high-level description of the proposed framework is presented in Algorithm 1 which consists of three phases: *Training, Evaluation* and *Weighted-Voting Prediction*. In the Training phase, the self-labeled algorithms, which constitute the ensemble are trained utilizing the same labeled L and unlabeled dataset U (Steps 1–3). Subsequently, in the Evaluation phase, the trained classifiers are evaluated using the training set L in order to calculate the weight matrix W (Steps 4–9). Finally, in the Weighted-Voting Prediction phase, the final hypothesis on each unlabeled example x of the test set combines the individual predictions of self-labeled algorithms utilizing the proposed weighted voting methodology (Steps 10–15). An overview of the proposed WvEnSL is depicted in Figure 1.

Algorithm 1: WvEnSL

Input:
- L – Set of labeled instances (Training labeled set).
- U – Set of unlabeled instances (Training unlabeled set).
- T – Set of unlabeled test instances (Testing set).
- D – Set of instances for evaluation (Evaluation set).
- $C = (C_1, C_2, \ldots, C_N)$ – Set of self-labeled classifiers which constitute the ensemble.

Output: The labels of instances in the testing set.

/* Phase I: Training */
Step 1: for $i = 1$ to N do
Step 2:   Train C_i using the labeled L and the unlabeled dataset U.
Step 3: end for

/* Phase II: Evaluation */
Step 4: for $i = 1$ to N do
Step 5:   Apply C_i on the evaluation set D.

Algorithm 1: *Cont.*

Step 6:   **for** $j = 1$ **to** M **do**
Step 7:   Calculate the weight

$$w_{i,j} = \frac{2p_j^{(C_i)}}{|D_j| + p_j^{(C_i)} + q_j^{(C_i)}}.$$

Step 8:   **end for**
Step 9: **end for**

/* Phase III: Weighted-Voting Prediction */
Step 10: **for each** $x \in T$ **do**
Step 11:   **for** $i = 1$ **to** N **do**
Step 12:   Apply classifier C_i on x.
Step 13:   **end for**
Step 14:   Predict the label $\hat{y}$ of x using

$$\hat{y} = \arg\max_j \sum_{i=1}^{N} w_{i,j} \chi_A(C_i(x) = j).$$

Step 15: **end for**

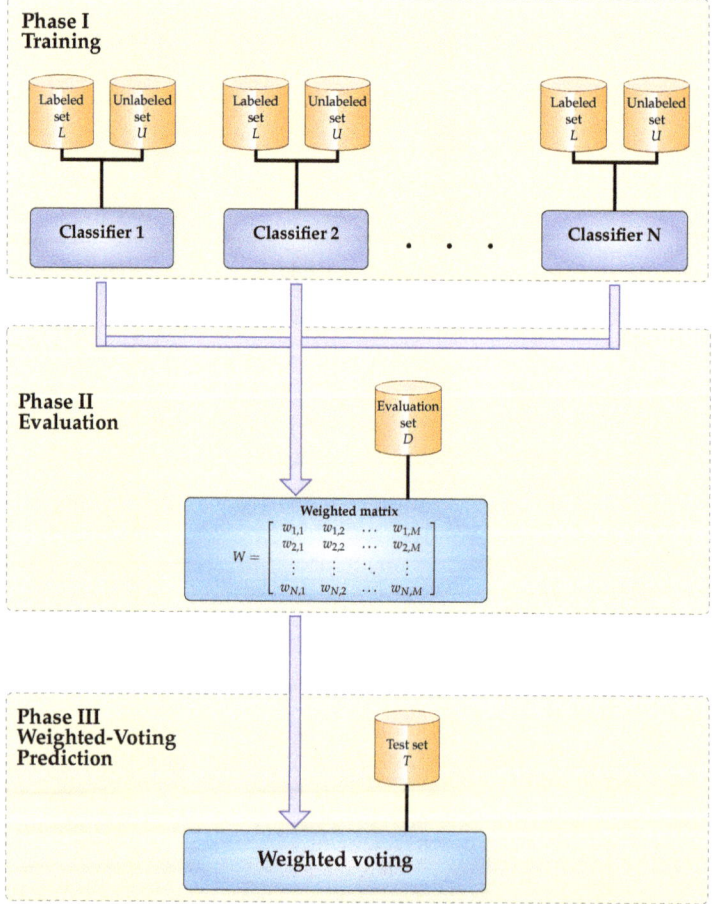

Figure 1. WvEnSL framework.

4. Experimental Methodology

In this section, we present a series of experiments in order to evaluate the performance of the proposed WvEnSL algorithm for X-ray classification against the most efficient ensemble self-labeled algorithms i.e., CST-Voting, DTCo and EnSL which utilize simple voting methodologies. The implementation code was written in JAVA, making use of the WEKA 3.9 Machine Learning Toolkit [27].

The performance of the classification algorithms is evaluated using the following performance metrics: F-measure (F_1) and Accuracy (Acc). It is worth mentioning that F_1 consists of a harmonic mean of precision and recall while Accuracy is the ratio of correct predictions of a classifier.

4.1. Datasets

The compared classification algorithms were evaluated utilizing the chest X-ray (Pneumonia) dataset, the Shenzhen lung mask (Tuberculosis) dataset and the CT Medical images dataset.

- *Chest X-ray (Pneumonia) dataset*: The dataset contains 5830 chest X-ray images (anterior-posterior) which were selected from retrospective cohorts of pediatric patients of one to five years old from Guangzhou Women and Children's Medical Center, Guangzhou. All chest X-ray imaging was performed as part of patients' routine clinical care. For the analysis of chest X-ray images, all chest radiographs were initially screened for quality control by removing all low quality or unreadable scans. The diagnoses for the images were then graded by two expert physicians before being cleared for training the artificial intelligence system. In order to account for any grading errors, the evaluation set was also checked by a third expert. The dataset was partitioned into two sets (training/testing). The training set consisting of 5216 examples (1341 normal, 3875 pneumonia) and the testing set with 624 examples (234 normal, 390 pneumonia) as in [28].
- *Shenzhen lung mask (Tuberculosis) dataset*: Shenzhen Hospital is one of the largest hospitals in China for infectious diseases with a focus both on their prevention, as well as treatment. The X-rays were collected within a one-month period, mostly in September 2012, as a part of the daily routine, using a Philips DR Digital Diagnost system. The dataset was constructed by manually-segmented lung masks for the Shenzhen Hospital X-ray set as presented in [29]. These segmented lung masks were originally utilized for the description of the lung segmentation technique in combination with lossless and lossy data augmentation. The segmentation masks for the Shenzhen Hospital X-ray set were manually prepared by students and teachers of the Computer Engineering Department, Faculty of Informatics and Computer Engineering, National Technical University of Ukraine "Igor Sikorsky Kyiv Polytechnic Institute" [29]. The set contained 279 normal CXRs and 287 abnormal ones with tuberculosis. All classification algorithms were evaluated using the stratified ten-fold cross-validation.
- *CT Medical images dataset*: This data collectioncontains 100 images [30] which constitute part of a much larger effort, focused on connecting cancer phenotypes to genotypes by providing clinical images matched to subjects from *the cancer genome Atlas* [31]. The images consist of the middle slice of all Computed Tomography (CT) images taken from 69 different patients. The dataset is designed to allow different methods to be evaluated for examining the trends in CT image data associated with using contrast and patient age. The basic idea is to identify image textures, statistical patterns and features correlating strongly with these traits and possibly build simple tools for automatically classifying these images when they have been misclassified (or finding outliers which could be suspicious cases, bad measurements, or poorly calibrated machines). All classification algorithms were evaluated using the stratified ten-fold cross-validation.

The training partition was randomly divided into labeled and unlabeled subsets. In order to study the influence of the amount of labeled data, four different ratios (R) of the training data were used: 10%, 20%, 30% and 40%. Using the recommendation established in [9,32] in the division process we do not maintain the class proportion in the labeled and unlabeled sets since the main aim of semi-supervised classification is to exploit unlabeled data for better classification results. Hence, we use a random selection of examples that will be marked as labeled instances, and the class label of the rest of the instances will be removed. Furthermore, we ensure that every class has at least one representative instance.

4.2. Performance Evaluation of WvEnSL against Ensemble Self-Labeled Algorithms

Next, we focus our interest on the experimental analysis for evaluating the classification performance of WvEnSL algorithm against the ensemble self-labeled algorithms CST-Voting and DTCo, which utilize simple voting methodologies. It is worth noticing that our main goal is to measure the effectiveness of the proposed weighted voting strategy over the simple majority voting; therefore, we will compare ensembles using identical set of classifiers. This will eliminate the source of discrepancy originated from unequal classifiers. Thus, the difference in accuracy can solely be attributed to the difference of voting methodologies.

Furthermore, the base learners utilized in all self-labeled algorithms are the Sequential Minimum Optimization (SMO) [33], the C4.5 decision tree algorithm [34] and the kNN algorithm [35] as in [2,7–9], which probably constitute the most effective and popular machine learning algorithms for classification problems [36].

- "CST-Voting (SMO)" stands for an ensemble of Co-training, Self-training and Tri-training with SMO as base learner using majority voting [2].
- "WvEnSL$_1$ (SMO)" stands for Algorithm WvEnSL using the same components classifiers as CST-Voting (SMO).
- "CST-Voting (C4.5)" stands for an ensemble of Co-training, Self-training and Tri-training with C4.5 as base learner using majority voting [2].
- "WvEnSL$_1$ (C4.5)" stands for Algorithm WvEnSL using the same components classifiers as CST-Voting (C4.5).
- "CST-Voting (kNN)" stands for an ensemble of Co-training, Self-training and Tri-training with kNN as base learner using majority voting [2].
- "WvEnSL$_1$ (kNN)" stands for Algorithm WvEnSL using the same components classifiers as CST-Voting (kNN).
- "DTCo" stands for an ensemble of Democratic-Co learning, Tri-training and Co-Bagging with C4.5 as base learner using majority voting [16].
- "WvEnSL$_2$" stands for Algorithm WvEnSL using the same components classifiers as DTCo.
- "EnSL" stands for an ensemble of Self-training, Democratic-Co learning, Tri-training and Co-Bagging with C4.5 as base learner using majority voting [17].
- "WvEnSL$_3$" stands for Algorithm WvEnSL using the same components classifiers as EnSL.

The configuration parameters for all supervised classifiers and self-labeled algorithms, utilized in our experiments, are presented in Table 1.

Table 1. Parameter specification for all the base learners and self-labeled methods used in the experimentation.

Algorithm		Parameters
SMO	Supervised base learner	$C = 1.0$, Tolerance parameter $= 0.001$, Pearson VII function-based kernel, Epsilon $= 1.0 \times 10^{-12}$, Fit logistic models $=$ true.
C4.5	Supervised base learner	Confidence level: $c = 0.25$, Minimum number of item-sets per leaf: $i = 2$, Prune after the tree building.
kNN	Supervised base learner	Number of neighbors $= 3$, Euclidean distance.
Self-training	Self-labeled (single classifier)	MaxIter $= 40$, $c = 95\%$.
Co-training	Self-labeled (multiple classifier)	MaxIter $= 40$, Initial unlabeled pool $= 75$
Tri-training	Self-labeled (multiple classifier)	No parameters specified.
Co-Bagging	Self-labeled (multiple classifier)	Committee members $= 3$, Ensemble learning $=$ Bagging.
Democratic-Co	Self-labeled (multiple classifier)	Classifiers $=$ kNN, C4.5, NB.
CST-Voting	Ensemble of self-labeled	No parameters specified.
DTCo	Ensemble of self-labeled	No parameters specified.
EnSL	Ensemble of self-labeled	No parameters specified.

Tables 2–4 presents the performance of all ensemble self-labeled methods on Pneumonia dataset, Tuberculosis dataset and CT Medical dataset, respectively. Notice that the highest classification performance for each ensemble of classifiers and performance metric is highlighted in bold. The aggregated results showed that the new weighted voting strategy exploits the individual predictions of each component classifier more efficiently than the simple voting schemes, illustrating better classification performance. WvEnSL$_3$ exhibits the best performance, reporting the highest F_1-score and accuracy, relative to all classification benchmarks and labeled ratio, followed by WvEnSL$_2$. In more detail, WvEnSL$_3$ demonstrates 82.53–83.49%, 69.79–71.73% and 69–77% classification accuracy for Pneumonia dataset, Tuberculosis dataset and CT Medical dataset, respectively; while WvEnSL$_2$ reports 81.89–83.17%, 69.79–71.55% and 67–77%, in the same situations.

The statistical comparison of several classification algorithms over multiple datasets is fundamental in the area of machine learning and it is usually performed by means of a statistical test [2,7–9]. Since our motivation stems from the fact that we are interested in evaluating the rejection of the hypothesis that all the algorithms perform equally well for a given level based on their classification accuracy and highlighting the existence of significant differences between our proposed algorithm and the classical self-labeled algorithms, we utilized the non-parametric Friedman Aligned Ranking (FAR) [37] test.

Table 2. Performance evaluation of WvEnSL against ensemble self-labeled algorithms for Pneumonia dataset.

Algorithm	Ratio = 10%		Ratio = 20%		Ratio = 30%		Ratio = 40%	
	F_1	Acc	F_1	Acc	F_1	Acc	F_1	Acc
CST-Voting (SMO)	83.08%	75.32%	83.26%	75.64%	83.39%	75.80%	83.39%	75.80%
WvEnSL$_1$ (SMO)	83.39%	75.80%	83.48%	75.96%	83.76%	76.44%	83.85%	76.60%
CST-Voting (C4.5)	85.52%	79.97%	85.85%	80.45%	86.68%	81.73%	86.58%	81.57%
WvEnSL$_1$ (C4.5)	85.65%	80.13%	86.08%	80.77%	86.78%	81.89%	86.92%	82.05%
CST-Voting (kNN)	82.91%	75.48%	83.09%	75.80%	83.15%	75.96%	83.73%	76.76%
WvEnSL$_1$ (kNN)	83.63%	76.60%	83.73%	76.76%	83.95%	77.08%	84.23%	77.56%
DTCo	86.79%	81.41%	87.21%	82.05%	87.21%	82.05%	87.74%	82.85%
WvEnSL$_2$	87.12%	81.89%	87.44%	82.37%	87.54%	82.53%	87.97%	83.17%
EnSL	87.19%	82.05%	86.92%	81.57%	87.34%	82.21%	87.61%	82.69%
WvEnSL$_3$	87.51%	82.53%	87.70%	82.69%	88.23%	83.49%	88.17%	83.49%

Table 3. Performance evaluation of WvEnSL against ensemble self-labeled algorithms for Tuberculosis dataset.

Algorithm	Ratio = 10%		Ratio = 20%		Ratio = 30%		Ratio = 40%	
	F_1	Acc	F_1	Acc	F_1	Acc	F_1	Acc
CST-Voting (SMO)	69.27%	69.43%	68.65%	68.37%	69.50%	69.61%	70.42%	70.32%
WvEnSL$_1$ (SMO)	69.73%	69.79%	69.73%	69.79%	70.32%	70.32%	71.00%	70.85%
CST-Voting (C4.5)	66.67%	67.31%	68.19%	68.02%	67.51%	68.20%	69.52%	69.79%
WvEnSL$_1$ (C4.5)	67.86%	68.20%	69.26%	69.26%	69.63%	69.79%	69.98%	70.14%
CST-Voting (kNN)	65.71%	66.08%	66.43%	66.96%	68.21%	68.55%	68.93%	69.26%
WvEnSL$_1$ (kNN)	65.83%	66.25%	67.14%	67.49%	68.57%	68.90%	69.40%	69.61%
DTCo	69.73%	69.79%	69.96%	69.96%	71.45%	71.20%	71.80%	71.55%
WvEnSL$_2$	69.73%	69.79%	70.19%	70.14%	71.58%	71.38%	71.80%	71.55%
EnSL	69.73%	69.79%	69.96%	69.96%	71.00%	70.85%	71.58%	71.38%
WvEnSL$_3$	69.73%	69.79%	70.19%	70.14%	71.58%	71.38%	72.03%	71.73%

Table 4. Performance evaluation of WvEnSL against ensemble self-labeled algorithms for CT Medical dataset.

Algorithm	Ratio = 10%		Ratio = 20%		Ratio = 30%		Ratio = 40%	
	F_1	Acc	F_1	Acc	F_1	Acc	F_1	Acc
CST-Voting (SMO)	66.67%	66.00%	70.00%	70.00%	73.08%	72.00%	75.00%	74.00%
WvEnSL$_1$ (SMO)	68.00%	68.00%	71.29%	71.00%	73.79%	73.00%	75.73%	75.00%
CST-Voting (C4.5)	67.96%	67.00%	71.84%	71.00%	73.79%	73.00%	73.79%	73.00%
WvEnSL$_1$ (C4.5)	69.90%	69.00%	73.79%	73.00%	75.00%	74.00%	75.73%	75.00%
CST-Voting (kNN)	66.00%	66.00%	69.90%	69.00%	73.79%	73.00%	73.27%	73.00%
WvEnSL$_1$ (kNN)	66.67%	67.00%	70.59%	70.00%	72.00%	72.00%	74.75%	75.00%
DTCo	66.02%	65.00%	69.90%	69.00%	72.55%	72.00%	74.29%	73.00%
WvEnSL$_2$	67.33%	67.00%	71.29%	71.00%	72.55%	72.00%	76.92%	76.00%
EnSL	64.08%	63.00%	71.84%	71.00%	74.29%	73.00%	74.29%	73.00%
WvEnSL$_3$	69.90%	69.00%	75.73%	75.00%	76.47%	76.00%	77.67%	77.00%

Let r_i^j be the rank of the j-th of k learning algorithms on the i-th of M problems. Under the null-hypothesis H_0, which states that all the algorithms are equivalent, the Friedman aligned ranks test statistic is defined by:

$$F_{AR} = \frac{(k-1)\left[\sum_{j=1}^{k} \hat{R}_j^2 - (kM^2/4)(kM+1)^2\right]}{\frac{kM(kM+1)(2kM+1)}{6} - \frac{1}{k}\sum_{i=1}^{M}\hat{R}_i^2}$$

where $\hat{R}_i$ is equal to the rank total of the i-th dataset and $\hat{R}_j$ is the rank total of the j-th algorithm. The test statistic F_{AR} is compared with the χ^2 distribution with $(k-1)$ degrees of freedom. It is worth noticing that, FAR test does not require the commensurability of the measures across different datasets, since it is non-parametric, neither assumes the normality of the sample means, and thus, it is robust to outliers.

Additionally, in order to identify which algorithms report significant differences, the Finner test [38] with a significance level $\alpha = 0.05$, is applied as a post-hoc procedure. More analytically, the Finner procedure adjusts the value of α in a step-down manner. Let $p_1, p_2, \ldots, p_{k-1}$ be the ordered p-values with $p_1 \leq p_2 \leq \cdots \leq p_{k-1}$ and $H_1, H_2, \ldots, H_{k-1}$ be the corresponding hypothesis. The Finner procedure rejects H_1–H_{i-1} if i is the smallest integer such that $p_i > 1 - (1-\alpha)^{(k-1)/i}$, while the adjusted Finner p-value is defined by:

$$p_F = \min\left\{1, \max\left\{1 - (1-p_j)^{(k-1)/j}\right\}\right\},$$

where p_j is the p-value obtained for the j-th hypothesis and $1 \leq j \leq i$. It is worth mentioning that the test rejects the hypothesis of equality when the p_F is less than α.

The control algorithm for the post-hoc test is determined by the best (lowest) ranking obtained in each FAR test. Moreover, the adjusted p-value with Finner's test (p_F) was presented based on the corresponding control algorithm at the α level of significance while the post-hoc test rejects the hypothesis of equality when the value of p_F is less than the value of a. It is worth mentioning that the FAR test and the Finner post-hoc test were performed based on the classification accuracy of each algorithm over all datasets and labeled ratio.

Table 5 presents the information of the statistical analysis performed by nonparametric multiple comparison procedures for all ensemble self-labeled algorithms. The interpretation of Table 5 demostrates that WvEnSL$_3$ reports the highest probability-based ranking by statistically presenting better results, followed by WvEnSL$_2$ and WvEnSL$_1$ (C4.5). Moreover, it is worth mentioning that all weighted voting ensemble outperformed the corresponding ensemble which utilize classical voting schemes. Finally, based on the statistical analysis, we can easily conclude that the new weighted voting scheme had a significant impact on the performance of all ensemble of self-labeled algorithms.

Table 5. Friedman Aligned Ranking (FAR) test and Finner post-hoc test.

Algorithm	FAR	Finner Post-Hoc Test	
		p_F-Value	Null Hypothesis
WvEnSL$_3$	15.667	-	-
WvEnSL$_2$	34.958	0.174312	accepted
WvEnSL$_1$ (C4.5)	44.208	0.049863	rejected
EnSL	47.958	0.029437	rejected
DTCo	51.125	0.018734	rejected
CST-Voting (C4.5)	64.042	0.001184	rejected
WvEnSL$_1$ (SMO)	71.417	0.000194	rejected
CST-Voting (SMO)	88.292	0.000001	rejected
WvEnSL$_1$ (kNN)	89.083	0.000001	rejected
CST-Voting (kNN)	98.250	0.000001	rejected

4.3. Performance Evaluation of WvEnSL against Classical Supervised Algorithms

Next, we compare the classification performance of the proposed algorithm against the classical supervised classification algorithms: SMO, C4.5 and kNN. Moreover, we compare the performance of iCST-Voting against the ensemble of classifiers (Voting) which combines the individual predictions of the supervised classifiers utilizing a simple majority voting strategy. It is worth noticing that

- we selected WvEnSL$_3$ from all versions of the proposed algorithm since it presented the best overall performance.
- all supervised algorithms were trained using with 100% of the training set while WvEnSL$_3$ was trained using $R = 40\%$ of the training set.

Table 6 presents the performance of the proposed algorithm WvEnSL$_3$ against the supervised algorithms SMO, C4.5, kNN and Voting on Pneumonia dataset, Tuberculosis dataset and CT Medical dataset. As above mentioned, the highest classification performance for each labeled ratio and performance metric is highlighted in bold. The aggregated results show that WvEnSL$_3$ is the most efficient algorithm since it illustrates the best overall classification performance. More specifically, WvEnSL$_3$ exhibits the highest F_1-score and classification accuracy on Pneumonia and Tuberculosis datasets, while for CT Medical dataset, WvEnSL$_3$ reports the second best performance, considerably outperformed by C4.5.

Table 6. Performance evaluation WvEnSL$_3$ against state-of-the-art supervised algorithms on Pneumonia dataset, Tuberculosis dataset and CT Medical dataset.

Algorithm	Pneumonia		Tuberculosis		CT Medical	
	F_1	Acc	F_1	Acc	F_1	Acc
SMO	74.03%	76.76%	71.41%	71.37%	74.91%	75.00%
C4.5	72.41%	74.83%	62.32%	62.36%	**79.82%**	**80.00%**
3NN	72.32%	74.51%	67.51%	67.49%	67.08%	67.00%
Voting	73.34%	76.12%	71.00%	71.02%	74.07%	74.00%
WvEnSL$_3$	**88.17%**	**83.49%**	**72.03%**	**71.73%**	77.67%	77.00%

5. Conclusions

In this work, we proposed a new weighted voting ensemble self-labeled algorithm for the detection of lung abnormalities from X-rays, entitled WvEnSL. The proposed algorithm combines the individual predictions of self-labeled algorithms utilizing a new weighted voting methodology. The significant advantage of WvEnSL is that weights assigned on each component classifier of the ensemble are based on its accuracy on each class of the dataset.

For testing purposes, the algorithm was extensively evaluated using the chest X-rays (Pneumonia) dataset, the Shenzhen lung mask (Tuberculosis) dataset and the CT Medical images dataset. Our numerical experiments indicated better classification accuracy of the WvEnSL and demonstrated the efficiency of the new weighted voting scheme, as statistically confirmed by the Friedman Aligned Ranks nonparametric test as well as the Finner post hoc test. Therefore, we can conclude that the new weighted voting strategy had a significant impact on the performance of all ensembles of self-labeled algorithms, exploiting the individual predictions of each component classifier more efficiently than the simple voting schemes. Finally, it is worth mentioning that efficient and powerful classification models could be developed by the adaptation of ensemble methodologies in the SSL framework.

In our future work, we intend to pursue extensive empirical experiments to compare the proposed WvEnSL with other algorithms belonging to different SSL classes, and evaluate its performance using various component self-labeled algorithms and base learners. Furthermore, since our preliminary numerical experiments are quite encouraging, our next step is to explore the performance of the proposed algorithm on imbalanced datasets [39,40] and incorporate our proposed methodology for multi-target problems [41–43]. Additionally, another interesting aspect is the use of other component classifiers in the ensemble and enhance our proposed framework with more sophisticated and theoretically sound criteria for the development of an advanced weighted voting strategy. Finally, we intend to investigate and evaluate different strategies for the selection of the evaluation set.

Author Contributions: I.E.L., A.K., V.T. and P.P. conceived of the idea, designed and performed the experiments, analyzed the results, drafted the initial manuscript and revised the final manuscript.

Funding: This research received no external funding.

Conflicts of Interest: The authors declare no conflict of interest.

References

1. Van Ginneken, B.; Stegmann, M.B.; Loog, M. Segmentation of anatomical structures in chest radiographs using supervised methods: A comparative study on a public database. *Medical Image Anal.* **2006**, *10*, 19–40. [CrossRef] [PubMed]
2. Livieris, I.; Kanavos, A.; Tampakas, V.; Pintelas, P. An ensemble SSL algorithm for efficient chest X-ray image classification. *J. Imaging* **2018**, *4*, 95. [CrossRef]
3. Zhu, X.; Goldberg, A. Introduction to semi-supervised learning. *Synth. Lect. Artif. Intell. Mach. Learn.* **2009**, *3*, 1–130. [CrossRef]
4. Chapelle, O.; Scholkopf, B.; Zien, A. Semi-supervised learning. *IEEE Trans. Neural Netw.* **2009**, *20*, 542–542. [CrossRef]
5. Levatic, J.; Dzeroski, S.; Supek, F.; Smuc, T. Semi-supervised learning for quantitative structure-activity modeling. *Informatica* **2013**, *37*, 173–179.
6. Levatić, J.; Ceci, M.; Kocev, D.; Džeroski, S. Semi-supervised classification trees. *J. Intell. Inf. Syst.* **2017**, *49*, 461–486. [CrossRef]
7. Livieris, I.; Kanavos, A.; Tampakas, V.; Pintelas, P. An auto-adjustable semi-supervised self-training algorithm. *Algorithm* **2018**, *11*, 139. [CrossRef]
8. Livieris, I.; Kiriakidou, N.; Kanavos, A.; Tampakas, V.; Pintelas, P. On ensemble SSL algorithms for credit scoring problem. *Informatics* **2018**, *5*, 40. [CrossRef]
9. Triguero, I.; García, S.; Herrera, F. Self-labeled techniques for semi-supervised learning: Taxonomy, software and empirical study. *Knowl. Inf. Syst.* **2015**, *42*, 245–284. [CrossRef]
10. Yarowsky, D. Unsupervised word sense disambiguation rivaling supervised methods. In Proceedings of the 33rd Annual Meeting of the Association For Computational Linguistics, Cambridge, MA, USA, 26–30 June 1995; pp. 189–196.
11. Blum, A.; Mitchell, T. Combining labeled and unlabeled data with co-training. In Proceedings of the 11th Annual Conference on Computational Learning Theory, Madison, WI, USA, 24–26 July 1998; pp. 92–100.
12. Zhou, Y.; Goldman, S. Democratic co-learning. In Proceedings of the 16th IEEE International Conference on Tools with Artificial Intelligence (ICTAI), Boca Raton, FL, USA, 15–17 November 2014; IEEE: Piscataway, NI, USA, 2004; pp. 594–602.

13. Zhou, Z.; Li, M. Tri-training: Exploiting unlabeled data using three classifiers. *IEEE Trans. Knowl. Data Eng.* **2005**, *17*, 1529–1541. [CrossRef]
14. Li, M.; Zhou, Z. Improve computer-aided diagnosis with machine learning techniques using undiagnosed samples. *IEEE Trans. Syst. Man Cybern. Part A Syst. Hum.* **2007**, *37*, 1088–1098. [CrossRef]
15. Hady, M.; Schwenker, F. Combining committee-based semi-supervised learning and active learning. *J. Comput. Sci. Technol.* **2010**, *25*, 681–698. [CrossRef]
16. Livieris, I.; Kotsilieris, T.; Anagnostopoulos, I.; Tampakas, V. DTCo: An ensemble SSL algorithm for X-rays classification. In *Advances in Experimental Medicine and Biology*; Springer: Berlin/Heidelberg, Germany, 2018.
17. Livieris, I.; Kanavos, A.; Pintelas, P. Detecting lung abnormalities from X-rays using and improved SSL algorithm. *Electron. Notes Theor. Comput. Sci.* **2019**, accepted for publication.
18. Livieris, I. A new ensemble self-labeled semi-supervised algorithm. *Informatica* **2018**, accepted for publication.
19. Jaeger, S.; Karargyris, A.; Candemir, S.; Folio, L.; Siegelman, J.; Callaghan, F.; Xue, Z.; Palaniappan, K.; Singh, R.; Antani, S.; et al. Automatic tuberculosis screening using chest radiographs. *IEEE Trans. Med. Imaging* **2014**, *33*, 233–245. [CrossRef]
20. Melendez, J.; van Ginneken, B.; Maduskar, P.; Philipsen, R.; Reither, K.; Breuninger, M.; Adetifa, I.; Maane, R.; Ayles, H.; Sánchez, C. A novel multiple-instance learning-based approach to computer-aided detection of tuberculosis on chest X-rays. *IEEE Trans. Med. Imaging* **2015**, *34*, 179–192. [CrossRef]
21. Alam, J.; Alam, S.; Hossan, A. Multi-Stage Lung Cancer Detection and Prediction Using Multi-class SVM Classifier. In Proceedings of the 2018 International Conference on Computer, Communication, Chemical, Material and Electronic Engineering, Rajshahi, Bangladesh, 8–9 February 2018; IEEE: Piscataway, NI, USA, 2018; pp. 1–4.
22. Madani, A.; Moradi, M.; Karargyris, A.; Syeda-Mahmood, T. Semi-supervised learning with generative adversarial networks for chest X-ray classification with ability of data domain adaptation. In Proceedings of the 15th IEEE International Symposium on Biomedical Imaging, Washington, DC, USA, 4–7 April 2018; IEEE: Piscataway, NI, USA, 2018; pp. 1038–1042.
23. Guan, Q.; Huang, Y. Multi-label chest X-ray image classification via category-wise residual attention learning. *Pattern Recognit. Lett.* **2018**, doi:10.1016/j.patrec.2018.10.027. [CrossRef]
24. Dietterich, T. Ensemble methods in machine learning. In *Multiple Classifier Systems*; Kittler, J., Roli, F., Eds.; Springer: Berlin/Heidelberg, Germany, 2001; Volume 1857, pp. 1–15.
25. Rokach, L. *Pattern Classification Using Ensemble Methods*; World Scientific Publishing Company: Singapore, 2010.
26. Powers, D.M. Evaluation: From precision, recall and F-measure to ROC, informedness, markedness and correlation. *J. Mach. Learn. Technol.* **2011**, *2*, 37–63.
27. Hall, M.; Frank, E.; Holmes, G.; Pfahringer, B.; Reutemann, P.; Witten, I. The WEKA data mining software: An update. *SIGKDD Explor. Newsl.* **2009**, *11*, 10–18. [CrossRef]
28. Kermany, D.; Goldbaum, M.; Cai, W.; Valentim, C.; Liang, H.; Baxter, S.; McKeown, A.; Yang, G.; Wu, X.; Yan, F. Identifying medical diagnoses and treatable diseases by image-based deep learning. *Cell* **2018**, *172*, 1122–1131. [CrossRef]
29. Stirenko, S.; Kochura, Y.; Alienin, O.; Rokovyi, O.; Gang, P.; Zeng, W.; Gordienko, Y. Chest X-ray analysis of tuberculosis by deep learning with segmentation and augmentation. *arXiv* **2018**, arXiv:1803.01199.
30. Albertina, B.; Watson, M.; Holback, C.; Jarosz, R.; Kirk, S.; Lee, Y.; Lemmerman, J. Radiology data from the cancer Genome Atlas Lung Adenocarcinoma [TCGA-LUAD] collection. *Cancer Imaging Arch.* **2016**.
31. Clark, K.; Vendt, B.; Smith, K.; Freymann, J.; Kirby, J.; Koppel, P.; Moore, S.; Phillips, S.; Maffitt, D.; Pringle, M. The Cancer Imaging Archive (TCIA): maintaining and operating a public information repository. *J. Digit. Imaging* **2013**, *26*, 1045–1057. [CrossRef]
32. Wang, Y.; Xu, X.; Zhao, H.; Hua, Z. Semi-supervised learning based on nearest neighbor rule and cut edges. *Knowl.-Based Syst.* **2010**, *23*, 547–554. [CrossRef]
33. Platt, J. *Advances in Kernel Methods—Support Vector Learning*; MIT Press: Cambridge, MA, USA, 1998.
34. Quinlan, J. *C4.5: Programs for Machine Learning*; Morgan Kaufmann: San Francisco, CA, USA, 1993.
35. Aha, D. *Lazy Learning*; Kluwer Academic Publishers: Dordrecht, The Netherlands, 1997.
36. Wu, X.; Kumar, V.; Quinlan, J.; Ghosh, J.; Yang, Q.; Motoda, H.; McLachlan, G.; Ng, A.; Liu, B.; Yu, P.; et al. Top 10 algorithms in data mining. *Knowl. Inf. Syst.* **2008**, *14*, 1–37. [CrossRef]

37. Hodges, J.; Lehmann, E. Rank methods for combination of independent experiments in analysis of variance. *Ann. Math. Stat.* **1962**, *33*, 482–497. [CrossRef]
38. Finner, H. On a monotonicity problem in step-down multiple test procedures. *J. Am. Stat. Assoc.* **1993**, *88*, 920–923. [CrossRef]
39. Li, S.; Wang, Z.; Zhou, G.; Lee, S. Semi-supervised learning for imbalanced sentiment classification. In Proceedings of the IJCAI Proceedings-International Joint Conference on Artificial Intelligence, Barcelona, Catalonia, Spain, 16–22 July 2011; Volume 22, p. 1826.
40. Jeni, L.A.; Cohn, J.F.; De La Torre, F. Facing imbalanced data—Recommendations for the use of performance metrics. In Proceedings of the Humaine Association Conference on Affective Computing and Intelligent Interaction, Geneva, Switzerland, 2–5 September 2013; IEEE: Piscataway, NI, USA, 2013; pp. 245–251.
41. Levatić, J.; Ceci, M.; Kocev, D.; Džeroski, S. Self-training for multi-target regression with tree ensembles. *Knowl.-Based Syst.* **2017**, *123*, 41–60. [CrossRef]
42. Levatić, J.; Kocev, D.; Džeroski, S. The importance of the label hierarchy in hierarchical multi-label classification. *J. Intell. Inf. Syst.* **2015**, *45*, 247–271. [CrossRef]
43. Levatić, J.; Kocev, D.; Ceci, M.; Džeroski, S. Semi-supervised trees for multi-target regression. *Inf. Sci.* **2018**, *450*, 109–127. [CrossRef]

© 2019 by the authors. Licensee MDPI, Basel, Switzerland. This article is an open access article distributed under the terms and conditions of the Creative Commons Attribution (CC BY) license (http://creativecommons.org/licenses/by/4.0/).

Article

Exploring an Ensemble of Methods that Combines Fuzzy Cognitive Maps and Neural Networks in Solving the Time Series Prediction Problem of Gas Consumption in Greece

Konstantinos I. Papageorgiou [1,*], Katarzyna Poczeta [2], Elpiniki Papageorgiou [3], Vassilis C. Gerogiannis [3,*] and George Stamoulis [1]

1. Department of Computer Science & Telecommunications, University of Thessaly, 35100 Lamia, Greece; georges@cs.uth.gr
2. Department of Information Systems, Kielce University of Technology, 25-541 Kielce, Poland; k.piotrowska@tu.kielce.pl
3. Faculty of Technology, University of Thessaly-Gaiopolis, 41500 Gaiopolis, Larissa, Greece; elpinikipapageorgiou@uth.gr
* Correspondence: konpapageorgiou@uth.gr (K.I.P.); vgerogian@uth.gr (V.C.G.)

Received: 30 September 2019; Accepted: 31 October 2019; Published: 6 November 2019

Abstract: This paper introduced a new ensemble learning approach, based on evolutionary fuzzy cognitive maps (FCMs), artificial neural networks (ANNs), and their hybrid structure (FCM-ANN), for time series prediction. The main aim of time series forecasting is to obtain reasonably accurate forecasts of future data from analyzing records of data. In the paper, we proposed an ensemble-based forecast combination methodology as an alternative approach to forecasting methods for time series prediction. The ensemble learning technique combines various learning algorithms, including SOGA (structure optimization genetic algorithm)-based FCMs, RCGA (real coded genetic algorithm)-based FCMs, efficient and adaptive ANNs architectures, and a hybrid structure of FCM-ANN, recently proposed for time series forecasting. All ensemble algorithms execute according to the one-step prediction regime. The particular forecast combination approach was specifically selected due to the advanced features of each ensemble component, where the findings of this work evinced the effectiveness of this approach, in terms of prediction accuracy, when compared against other well-known, independent forecasting approaches, such as ANNs or FCMs, and the long short-term memory (LSTM) algorithm as well. The suggested ensemble learning approach was applied to three distribution points that compose the natural gas grid of a Greek region. For the evaluation of the proposed approach, a real-time series dataset for natural gas prediction was used. We also provided a detailed discussion on the performance of the individual predictors, the ensemble predictors, and their combination through two well-known ensemble methods (the average and the error-based) that are characterized in the literature as particularly accurate and effective. The prediction results showed the efficacy of the proposed ensemble learning approach, and the comparative analysis demonstrated enough evidence that the approach could be used effectively to conduct forecasting based on multivariate time series.

Keywords: fuzzy cognitive maps; neural networks; time series forecasting; ensemble learning; prediction; machine learning; natural gas

1. Introduction

Time series forecasting is a highly important and dynamic research domain, which has wide applicability to many diverse scientific fields, ranging from ecological modeling to energy [1], finance [2,3], tourism [4,5], and electricity load [6,7]. A summary of applications regarding forecasting in various areas can be found in a plethora of review papers published in the relevant literature [8–15]. One of the main challenges in the field of time series forecasting is to obtain reasonable and accurate forecasts of future data from analyzing previous historical data [16]. Following the literature, numerous research studies show that forecasting accuracy is improved when different models are combined, while the resulting model seems to outperform all component models [16]. Thus, the combination of forecasts from different models or algorithms becomes a promising field in the prediction of future data.

Researchers can choose between linear and nonlinear time series forecasting methods, depending on the nature of the model they are working on. Linear methods, like autoregressive moving average (ARMA) and autoregressive integrated moving average (ARIMA) [17] are among the methodologies which have found wide applicability in real-world applications. Traffic [18], energy [19–21], economy [22], tourism [23], and health [24] are some example fields in which the ARIMA forecasting technique was used. Considering though that most real-world problems are characterized by a non-linear behavior, many researchers have investigated the use of non-linear techniques for times series-based forecasting and prediction. In particular, machine learning and other intelligent methods have been chosen to address possible nonlinearities in time series modeling, such as nonlinear ARMA time series models, needing though to tackle the type of nonlinearity that is not usually known. Along with the computing power growth and the evolution of data management techniques, there has been a growing interest in the use of advanced artificial intelligence technologies, like artificial neural networks (ANNs) [25] and fuzzy logic systems for forecasting purposes. ANNs and fuzzy logic systems use more sophisticated generic model structures having the ability to incorporate the characteristics of complex data and produce accurate time series models [26], while they also incorporate the advantageous features of nonlinear modeling and data-based learning capabilities. Moreover, unlike traditional time series methods that are not able to adequately capture nonlinear patterns of data, neural networks are usually involved in predicting consumption demand during periods of low or extremely high demand [27]. Among all types of ANN models, the feed-forward network model with backpropagation training procedure (FFN-BP) is one of the most commonly used approaches [28].

The technique of combining the predictions of multiple classifiers to produce a single classifier is referred to as an ensemble [29–32]. Ensemble building has recently attracted much attention among researchers as an effective method that improves the performance of the resulting model for performing classification and regression tasks. It has been demonstrated that an ensemble of individual predictors performs better than a single predictor in the average [33,34], thus achieving better prediction accuracy than that of the individual predictors. Two popular methods for creating accurate ensembles are bagging [29] and boosting [35,36]. There are not many research works concerning potency forecasting of a model in advance, as this is often a difficult task. Trying to avoid the risk of combining models that have poor performance regarding prediction may result in an overall model with deteriorated forecasting accuracy. An ensemble model, however, should rather be formed solely from component models that are rated as adequate, if not good enough, for their forecasting capabilities [37]. Usually, the single neural network (NN) models are combined to create NNs ensembles, to tackle sampling and modeling uncertainties that could probably weaken the forecasting accuracy and robustness of component NN models. Since individual component model could be sensitive under different circumstances, ensembling them results in more powerful outcomes in the context of a decision-making process.

1.1. Related Literature

In recent years, the natural gas market has progressed greatly in terms of fast development and, thus, it has become a very competitive area. Undoubtedly, natural gas is among primary energy sources and has a significant environmental role considering its valuable environmental benefits, such as low-level emissions of greenhouse gases in comparison with other non-renewable energy sources [38,39]. Natural gas demand seems to increase considerably due to several socio-economic and political reasons, while the price and environmental concerns are significant regulatory factors affecting natural gas demand. Therefore, the prediction of natural gas consumption, as a time series forecasting problem, is becoming important in contemporary energy systems, allowing energy policymakers to apply effective strategies to guarantee sufficient natural gas supplies.

So far, many research papers have tried to give a clear insight regarding natural gas forecasting by suggesting models for predicting the consumption of this non-renewable energy source. A summary of natural gas consumption forecasting regarding prediction methods, input variables used for modeling, as well as prediction area, can be found in many review papers in the relevant literature. Particularly, there is a thorough literature survey of published papers [38–40] that classifies various models and techniques that have been recently applied in the field of natural gas forecasting, with respect to the paradigm that each model/technique is based on, while there has been also an attempt by researchers to classify all models applied in this area according to their performance characteristics, as well as to offer some future research directions. The models presented were developed by researchers to predict natural gas consumption on an hourly, daily, weekly, monthly, or yearly basis, in an attempt to predict natural gas consumption with an acceptable degree of accuracy. Accurate forecasting of natural gas consumption can be particularly important for project planning, engineering design, pipeline operation, gas imports, tariff design, and optimal scheduling of a natural gas supply system [41].

Due to the need for distribution planning, especially in residential areas, the increasing demand for natural gas, and the restricted natural gas network in many countries, the forecasting of consumption on a daily and weekly basis seems to be of high importance. In the relevant literature, many suggestions apply various ANN topologies and methods to support day-ahead natural gas demand prediction [42–50]. For example, [42] used ANN, while [45] used a combination of ANN forecasters for predicting gas consumption at a citywide distribution level. For the same distribution level (citywide), a strategy was proposed in [51] to estimate the forecasting risk by using hourly consumption data. Having as aim to find the best solution for natural gas consumption, the researchers in [52] used linear regression with the sliding window technique, while in [53], a univariate artificial bee colony-based artificial neural network (ANN-ABC) was applied to minimize error in the case of forecasting day-ahead demand. The researchers in [54] also considered various methods for the prediction of daily gas consumption, such as the seasonal autoregressive integrated moving average model with exogenous inputs (SARIMAX), multi-layer perceptron ANN (ANN-MLP), ANN with radial basis functions (ANN-RBF), and multivariate ordinary least squares (OLS).

In the relevant literature, there are also some research works on neuro-fuzzy methods and genetic algorithms applied in natural gas demand, as in [41,55–58]. Specifically, a novel hybrid model that combines the wavelet transform (WT), genetic algorithm (GA), adaptive neuro-fuzzy inference system (ANFIS), and feed-forward neural network (FFNN) was recently examined in [41] and applied to the Greek natural gas grid. Moreover, evolutionary fuzzy cognitive maps (FCMs) were recently used for time series problems modeling and forecasting. FCMs can be understood as recurrent neural networks inheriting many features from them, such as learning capabilities, which elevate the performance of FCMs in modeling and prediction and further helped FCMs to gain momentum over recent years [59,60]. The researchers in [61,62] were the first to examine the application of FCMs to time series modeling, proposing nodes selection criteria in an FCM, which was used to model univariate time series. Further techniques for simplifying FCMs by removing nodes and weights were investigated, while a dynamic optimization of the FCM structure was studied in [63] for univariate time series forecasting. Concerning multivariate interval-valued time series, an evolutionary algorithm for learning fuzzy grey cognitive maps was developed as a nonlinear predictive model [64]. Taking one step further, the researchers in [65] and [66] enhanced the evolutionary FCMs with the structure optimization genetic algorithm (SOGA). These approaches can be used to automatically construct an FCM model after selecting the crucial concepts and defining the relationships between them by taking into consideration any available historical data. An example regarding rented bikes' count prediction was examined, where SOGA-FCM was compared with the multi-step gradient method (MGM) [67] and the real-coded genetic algorithm (RCGA) [68]. A two-stage prediction model for multivariate time series prediction, based on the efficient capabilities of evolutionary fuzzy cognitive maps (FCMs) and enhanced by structure optimization algorithms and artificial neural networks (ANNs), was introduced in [69]. Furthermore, the researchers in [21,60] recently conducted a preliminary study on implementing FCMs with NNs for natural gas prediction.

1.2. Research Aim and Approach

The purpose of this paper was to propose a new forecast combination approach resulting from FCMs, ANNs, and hybrid models. This ensemble forecasting method, including the two most popular ensemble methods, the Average and the Error-based, is based on ANNs, FCMs with learning capabilities, as well as on a hybrid FCM-ANN model with different configurations, to produce an accurate non-linear time series model for the prediction of natural gas consumption. A real case study problem of natural gas consumption in Greece was performed to show the applicability of the proposed approach. Furthermore, in order to validate the proposed forecasting combination approach, a comparison analysis between the ensemble methods and an innovative machine learning technique, the long short-term memory (LSTM) algorithm (which is devoted to time series forecasting), was conducted, and the results demonstrated enough evidence that the proposed approach could be used effectively to conduct forecasting based on multivariate time series. The LSTM algorithm, as an advanced recurrent NN method, was previously used for short-term natural gas demand forecasting in Greece [70]. In that research paper, LSTM was applied in one day-ahead natural gas consumption, forecasting for the same three Greek cities, which were also examined in the case study presented in the current paper. Many similar works can be found in the literature that examine various forecast combinations in terms of accuracy and error variability but, in the present work, an innovative approach that combines FCMs, ANNS, and hybrid FCM-ANN models, producing a non-linear time series model for the prediction of natural gas consumption, was studied exclusively, contributing to the novelty of the current study. The results demonstrated in a clear way that the proposed approach had attained better accuracies than other individual models. This study justified the superiority of the selective ensemble method over combining the important features and capabilities of the models that consist of the overall approach, making it a useful tool for future work.

The outline of the paper is as follows. Section 2 describes the material and methods of our research study; Section 2.1 describes the case study problem and refers to the datasets of natural gas demand that are used, whereas Section 2.2 presents the analyzed approaches for time series forecasting based on ANNs, FCMs with evolutionary learning algorithms, and their hybrid combinations. The most widely used ensemble methods for forecasting problems (i.e., the error-based and the simple average method) are also presented in Section 2.2. In Section 3, the proposed forecasting combination approach is described. The same Section presents the evaluation criteria, which we have used to analyze the performance of the analyzed approaches for natural gas prediction. Section 4 presents the results of simulation analysis for three different Greek cities, as well as the conducted comparative analysis of the proposed approach with other intelligent techniques. A discussion of the results highlights the main findings of the proposed ensemble forecasts approach. Section 5 summarizes the main conclusions of the paper with further discussion and suggestions about future research expansion.

2. Materials and Methods

2.1. Material-Dataset

In the considered case study, three different prediction datasets of natural gas demand, derived from different districts in Greece, were analyzed from the records of the Hellenic Gas Transmission System Operator S.A. (www.desfa.gr, DESFA). DESFA company is responsible for the operation, management, exploitation, and development of the Greek Natural Gas System and its interconnections in a technically sound and economically viable way. From 2008, DESFA provides historical data of transmission system operation and natural gas deliveries/off-takes. In this research work, historical data with the values of gas consumption for a period of five years, from 2013 to 2017, were used as initial data to accomplish forecasting. These data were split into training and testing data, where usually the training data came from the first four years and were used for learning models, whereas the data of the last year were used for testing the applied artificial intelligence models.

It is crucial for an efficient forecast to properly select the number and types of inputs. Thus, we emphasized on defining proper input candidates. Six different inputs for time series prediction were considered. The first three inputs were devoted to month indicator, day indicator, and mean temperature. Specifically, concerning the calendar indicators, we used one input for months and one input for days coding. Let $m = 1, 2, \ldots, 12$ be the number of months. We considered the following matching: January/1, February/2, ... , December/12. Let $l = 1, 2, \ldots, 7$ be the number of days. The day type matching was as follows: Monday/1, Tuesday/2, ... , Sunday/7. The temperature data were obtained by the nearest to the distribution gas point station. The rest three inputs were the previously measured values of natural gas demand, for one-day before, two-day before, and the current day. These six variables were used to form the input pattern of the FCM. The output referred to the total daily demand for the specific distribution point.

The features that were gathered and used in our study to form the FCM model were enough and properly selected according to the relevant literature. From a recent literature review regarding the prediction of natural gas consumption [40], it can be seen that past gas consumption combined with meteorological data (especially temperature) are the most commonly used input variables for the prediction of natural gas consumption. A recent study [41] used past consumption, temperature, months, and days of the week, while in [55], day of week and demand of the same day in the previous year were used as input variables for natural gas forecasting. Considering the above practices described in the literature, it can be concluded that the features used in the current work were enough to predict the consumption of natural gas for the selected areas.

The Greek cities of Thessaloniki, Athens, and Larissa were selected for the conducted simulation analysis and comparison of the best performing algorithms. These different natural gas consumption datasets may offer insight into whether the analyzed algorithms perform equally in different locations, where the energy demand could be completely different for the same days.

2.2. Methods

2.2.1. Fuzzy Cognitive Maps Overview

A fuzzy cognitive map (FCM) is a directed graph in which nodes denote concepts important for the analyzed problem, and links represent the causal relationships between concepts [71]. It is an effective tool for modeling decision support systems. FCMs have been applied in many research domains, e.g., in business performance analysis [72], strategy planning [73], modeling virtual worlds [74], time series prediction [69], and adoption of educational software [75].

The FCM model can be used to perform simulations by utilizing its dynamic model. The values of the concepts change in time as simulation goes on [68]. The new values of the concepts can be calculated based on the popular dynamic model described as follows [59]:

$$X_i(t+1) = F\left(X_i(t) + \sum_{\substack{j=1 \\ j \neq i}}^{n} X_j(t) \cdot w_{j,i}\right) \quad (1)$$

where $X_i(t)$ is the value of the ith concept at the tth iteration, $w_{j,i}$ is the weight of the connection (relationship) between the jth concept and the ith concept, t is discrete-time, $i.j = 1, 2, \ldots, n$, n is the number of concepts, $F(x)$ is the sigmoidal transformation function [58]:

$$F(x) = \frac{1}{1+e^{-cx}} \qquad (2)$$

where c is a parameter, $c > 0$. The weights of the relationships show how causal concepts affect one another. If $w_{j,i} > 0$, then an increase/decrease in the value of the jth concept will increase/decrease the value of the ith concept. If $w_{j,i} < 0$, then an increase/decrease in the value of the jth concept will decrease/increase the value of the ith concept. If $w_{j,i} = 0$, there is no causal relationship between the jth and the ith concepts [74].

The FCM structure is often constructed based on expert knowledge or surveys [74]. We could also use machine learning algorithms and available historical data to construct the FCM model and determine the weights of the relationships between the FCM's concepts.

2.2.2. Fuzzy Cognitive Maps Evolutionary Learning

Evolutionary algorithms are popular techniques for FCMs learning. In this paper, we explored two effective techniques: the real-coded genetic algorithm (RCGA) [68] and the structure optimization genetic algorithm (SOGA) [69].

Real-Coded Genetic Algorithm (RCGA)

The RCGA algorithm defines individual in the population as follows [24]:

$$W' = \left[w_{1,2}, w_{1,3}, w_{1,4}, \ldots, w_{j,i}, \ldots, w_{n,n-1}\right]^T \qquad (3)$$

where $w_{j,i}$ is the weight of the relationship between the jth concept and the ith concept.

Individual in the population is evaluated with the use of a fitness function based on data error [66]:

$$fitness_p(MSE_{tr}(l)) = \frac{1}{aMSE_{tr}(l)+1} \qquad (4)$$

where a is a parameter, l is the number of generation, $l = 1, \ldots, L$, L is the maximum number of generations, p is the number of individual, $p = 1, \ldots, P$, P is the population size, and $MSE_{tr}(l)$ is the data error, described as follows:

$$MSE_{tr}(l) = \frac{1}{N_{tr}} \sum_{t=1}^{N_{tr}} e_t^2 \qquad (5)$$

where $t = 1, \ldots, N_{tr}$, N_{tr} is the number of training records, and e_t is the one-step-ahead prediction error at the tth iteration, described as follows:

$$e_t = Z(t) - X(t) \qquad (6)$$

where $X(t)$ is the predicted value of the output concept, and $Z(t)$ is the desired value of the output concept.

When the maximum number of generations L is reached, or the condition (7) is met, which means that the learning process is successful, then the RCGA stops.

$$fitness_p(MSE_{tr}(l)) > fitness_{max} \qquad (7)$$

where $fitness_p(MSE_{tr}(l))$ is the fitness function value for the best individual, and $fitness_{max}$ is a parameter.

Structure Optimization Genetic Algorithm (SOGA)

The SOGA algorithm is an extension of the RCGA algorithm [65,66] that allows the decision-maker to determine the most significant concepts and the relationships between them.

Individual is evaluated based on the fitness function based on new data error, described as follows [66]:

$$MSE'_{tr}(l) = MSE_{tr}(l) + b_1 \frac{n_r}{n^2} MSE_{tr}(l) + b_2 \frac{n_c}{n} MSE_{tr}(l) \qquad (8)$$

where b_1, b_2 are the parameters of the fitness function, n_r is the number of the non-zero relationships, n_c is the number of the concepts in the analyzed model, n is the number of all possible concepts, l is the number of generation, $l = 1, \ldots, L$, L is the maximum number of generations.

The fitness function that follows (9) calculates the quality of each population.

$$fitness_p(MSE'_{tr}(l)) = \frac{1}{aMSE'_{tr}(l) + 1} \qquad (9)$$

where a is an experimentally defined parameter, p is the number of the individual, $p = 1, \ldots, P$, P is the population size, and $MSE'_{tr}(l)$ is the new error measure.

We could construct a less complex time series prediction model by removing the redundant concepts and connections between them with the use of a binary vector C and the proposed error function.

The algorithmic steps of the learning and analysis of the FCM in modeling prediction systems with the use of population-based algorithms (SOGA and RCGA) were analytically presented in [69].

For our experiments, the evolutionary operators, a) ranking selection, b) uniform crossover, and c) random mutation were used [76,77]. In addition, we applied elite strategy selection, while a probability of crossover P_c and mutation P_m was assigned to each population.

2.2.3. Artificial Neural Networks

An artificial neural network (ANN) is a collection of artificial neurons organized in the form of layers [25]. Neurons are connected by weighted connections to form a NN. The most widely used ANNs in time series prediction are the multilayer perceptrons with an input layer, an output layer, and a single hidden layer that lies between the input and output layer. The most common structure is an ANN that uses one or two hidden layers, as a feed-forward neural network with one hidden layer is able to approximate any continuous function. Supervised learning algorithms and historical data can be used for the learning process of ANNs. The output of each neuron can be calculated based on the following formula:

$$X(t) = F\left(\sum_{j=1}^{m} X_j(t) \cdot w_j + b\right) \qquad (10)$$

where $X_j(t)$ is the value of the jth input signal, $t = 1, \ldots, N_{tr}$, N_{tr} is the number of training records, w_j is the synaptic weight, m is the number of input signals, b is the bias, and F is the sigmoid activation function. Training a neural network needs the values of the connection weights and the biases of the neurons to be determined. There are many neural network learning algorithms. The most popular algorithm for ANN learning is the back-propagation method. In this learning method, the weights change their values according to the learning records until one epoch (an entire learning dataset) is reached. This method aims to minimize the error function, described as follows [14,78,79]:

$$MSE_{tr}(l) = \frac{1}{2N_{tr}} \sum_{t=1}^{N_{tr}} e_t^2 \qquad (11)$$

where $t = 1, \ldots, N_{tr}$, N_{tr} is the number of training records, l is the number of epoch, $l = 1, \ldots, L$, L is the maximum number of epochs, and e_t is the one-step-ahead prediction error at the tth iteration, which is equal to:

$$e_t = Z(t) - X(t) \qquad (12)$$

where $X(t)$ is the output value of the ANN, and $Z(t)$ is the desired value.

The modification of the weights in the back-propagation algorithm can be calculated by the formula:

$$\Delta_{w_{kj}}(l) = -\gamma \frac{\partial J(l)}{\partial w_{kj}(l)} \tag{13}$$

where $\Delta_{w_{kj}}(l)$ is a change of the weight w_{kj} at the *l*th epoch, γ is a learning coefficient.

Backpropagation algorithm with momentum modifies the weights according to the formula:

$$\Delta_{w_{kj}}(l) = -\gamma \frac{\partial J(l)}{\partial w_{kj}(l)} + \alpha \Delta_{w_{kj}}(l-1) \tag{14}$$

where α is a momentum parameter.

2.2.4. Hybrid Approach Based on FCMs, SOGA, and ANNs

The hybrid approach for time series prediction is based on FCMs, the SOGA algorithm, and ANNs [68]. This approach consists of two stages:

1. Construction of the FCM model based on the SOGA algorithm to reduce the concepts that have no significant influence on data error.
2. Considering the selected concepts (data attributes) as the inputs for the ANN and ANN learning with the use of backpropagation method with momentum.

This hybrid structure allows the decision-maker to select the most significant concepts for an FCM model using the SOGA algorithm. These concepts are used as inputs for the ANN model. Such a hybrid approach aims to find the most accurate model for time series prediction problems.

2.2.5. The Ensemble Forecasting Method

The most intuitive and popular way of forecast aggregation is to linearly combine the constituent forecasts [80]. There are various methods proposed in the literature for selecting the combining weights [81]. The most popular and widely used ensemble methods are the error-based and the simple average [82]. The easiest among them is the simple average in which all forecasts are weighted equally, often remarkably improving overall forecasting accuracy [82,83].

Considering that $Y = [y_1, y_2, y_3, \ldots, y_N]^T$ is the actual out-of-sample testing dataset of a time series and $\hat{Y}^i = [\hat{y}_1^i, \hat{y}_2^i, \ldots, \hat{y}_n^i]^T$ is the forecast for the i_{th} model, the linear combination of n forecasts is produced by [15]:

$$\hat{y}_k = w_1 \hat{y}_k^{(1)} + w_2 \hat{y}_k^{(2)} + \ldots + w_n \hat{y}_k^{(n)} = \sum_{i=1}^{n} w_i \hat{y}_k^{(i)}, \ \forall k = 1, 2, \ldots, N \tag{15}$$

Here, our analysis is based on these most popular ensemble methods. A brief discussion follows for each one.

- The *simple average (AVG)* method [82] is an unambiguous technique, which assigns the same weight to every single forecast. Based on empirical studies in the literature, it has been observed that the AVG method is robust and able to generate reliable predictions, while it can be characterized as remarkably accurate and impartial. Being applied in several models, with respect to effectiveness, the AVG improved the average accuracy when increasing the number of combined single methods [82]. Comparing the referent method with the weighted combination techniques, in terms of forecasting performance, the researchers in [84] concluded that a simple average combination might be more robust than weighted average combinations. In the simple average combination, the weights can be specified as follows:

$$w_i = \frac{1}{n}, \quad \forall i = 1, 2, \ldots, n \qquad (16)$$

- The *error-based (EB)* method [16] consists of component forecasts, which are given weights that are inversely proportional to their in-sample forecasting errors. For instance, researchers may give a higher weight to a model with lower error, while they may assign a less weight value to a model that presents more error, respectively. In most of the cases, the forecasting error is calculated using total absolute error statistic, such as the sum of squared error (SSE) [80,83]. The combining weight for individual prediction is mathematically given by:

$$w_i = e_i^{-1} / \sum_{i=1}^{n} e_i^{-1}, \quad \forall i = 1, 2, \ldots, n \qquad (17)$$

3. The Proposed Forecast Combination Methodology

In the rest of the paper, we explored a new advanced forecasting approach by introducing a different split of dataset in the case of daily, weekly, or monthly forecasting, as well as a combination of forecasts from multiple structurally different models, like ANN and FCM with various efficient learning algorithms and hybrid configurations of them. Also, the two most popular and usually used ensemble methods, the AVG and the EB methods, were applied to the ensemble forecasts to improve the prediction accuracy.

In the described ensemble scheme, the selection of the appropriate validation set, i.e., the selection of the parameter N_{vd} and the group size N_{tr}, is very important. The validation set should reflect the characteristics of the testing dataset that is practically unknown in advance. As such, in this study, we set the following process of data split. The data split takes place by removing 15% of the total dataset N and saving for later use as testing data. The remaining 85% of the dataset is then split again into an 82/18 ratio, resulting in the following portions: 70% for training and 15% for validation. Also, the group size N_{tr} (i.e., the training data) should be appropriately selected so that it is neither too small nor too large.

Due to the problem nature, as we work with time-series data, the most efficient method for resampling is the boosting/bootstrapping method [85]. In boosting, resampling is strategically geared to provide the most informative training data for each consecutive predictor. Therefore, in this study, an appropriate bootstrapping method was applied, so that the training dataset should have the same size at each resampling set, and the validation and testing sets should keep the same size (after excluding the k-values from the in-sample dataset).

The proposed effective forecast combination methodology for time series forecasting, presented in the paper, includes three main processing steps: data pre-processing to handle missing values, normalize the collected time-series data, and split the dataset; the various forecasting methods of ANNs, RCGA-FCMs, SOGA-FCMs, and hybrid SOGA FCM-ANN with their ensembles; and evaluation of the prediction results, implementing the two most popular and used ensemble methods of simple average (AVG) and error-based (EB). Figure 1 visually illustrates the suggested methodology.

In the followed approach, data preprocessing included outlier detection and removal, handling missing data, and data normalization, all of which were in accordance with the principles of Data Science practices described in corresponding literature. For outlier detection, the Z-score was first calculated for each sample on the data set (using the standard deviation value that is presented in the descriptive statistics Tables A1 and A2 in Appendix A). Then, a threshold was specified, and the data points that lied beyond this threshold were classified as outliers and were removed. Mean imputation was performed to handle missing values. Specifically, for numerical features, missing values were replaced by the mean feature value.

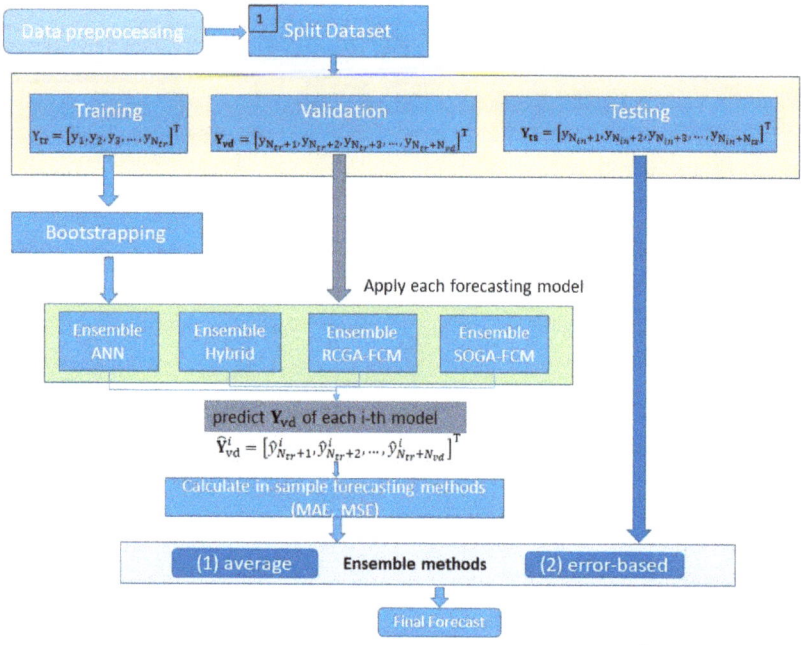

Figure 1. Flowchart of the proposed forecasting combination approach.

Each dataset was normalized to [0,1] before the forecasting models were applied. The normalized datasets were taking again their original values, while the testing phase was implemented. The data normalizations were carried out mathematically, as follows:

$$y_i^{(new)} = \frac{y_i - y^{(min)}}{y^{(max)} - y^{(min)}}, \forall i = 1, 2, \ldots, N \tag{18}$$

where $Y = [y_1, y_2, y_3, \ldots, y_{N_{train}}]^T$ is the training dataset, and $Y^{(new)} = [y_1^{(new)}, y_1^{(new)}, \ldots, y_N^{(new)}]^T$ is the normalized dataset, $y^{(min)}$ and $y^{(max)}$ are, respectively, the minimum and maximum values of the training dataset Y.

We selected the Min-Max normalization method [86] as it is one of the most popular and comprehensible methods, in terms of performance of the examined systems, while several researchers showed that it produces better (if not equally good) results with high accuracy, compared to the other normalization methods [87,88]. In [88], the Min-Max was valued as the second-best normalization method in the backpropagation NN model, justifying our choice to deploy this method for data normalization. Moreover, since the FCM concepts use values within the range [0,1] for the conducted simulations and do not deal with real values, the selected method seemed to be proper for our study. Also, this normalization approach was previously used in [66,69].

Due to our intention to suggest a generic forecasting combination approach (with ANNs, FCMs, and their hybrid structures) able to be applied in any time series dataset, the following steps are thoroughly presented and executed.

Step 1. (Split Dataset) We divided the original time series $Y = [y_1, y_2, y_3, \ldots, y_N]^T$ into the in-sample training dataset $Y_{tr} = [y_1, y_2, y_3, \ldots, y_{N_{tr}}]^T$, the in-sample validation dataset $Y_{vd} = [y_{N_{tr}+1}, y_{N_{tr}+2}, y_{N_{tr}+3}, \ldots, y_{N_{tr}+N_{vd}}]^T$, and the out-of-sample testing dataset

$Y_{ts} = \left[y_{N_{in}+1}, y_{N_{in}+2}, y_{N_{in}+3}, \dots, y_{N_{in}+N_{ts}} \right]^T$, so that $N_{in} = N_{tr} + N_{vd}$ is the size of the total in-sample dataset and $N_{in} + N_{ts} = N$, where N is the number of days, or weeks, or months, according to the short- or long-term prediction based on the time series horizon.

Step 2. (**Resampling method/Bootstrapping**). Let's consider k sets as training sets from the whole dataset every time. For example, in the monthly forecasting, we excluded one month every time from the initial in-sample dataset, starting from the first month of the time series values, and proceeding with next month till $k = 12$, (i.e., this means that 1 to 12 months were excluded from the initial in-sample dataset). Therefore, k subsets of training data were created and used for training. The remaining values of the in-sample dataset were used for validation, whereas the testing set remained the same. Figure 2 shows an example of this bootstrapping method for the ensemble SOGA-FCM approach. In particular, Figure 2a represents the individual forecasters' prediction values and their average error calculation, whereas, in Figure 2b, the proposed forecasting combination approach for SOGA-FCM is depicted for both ensemble methods.

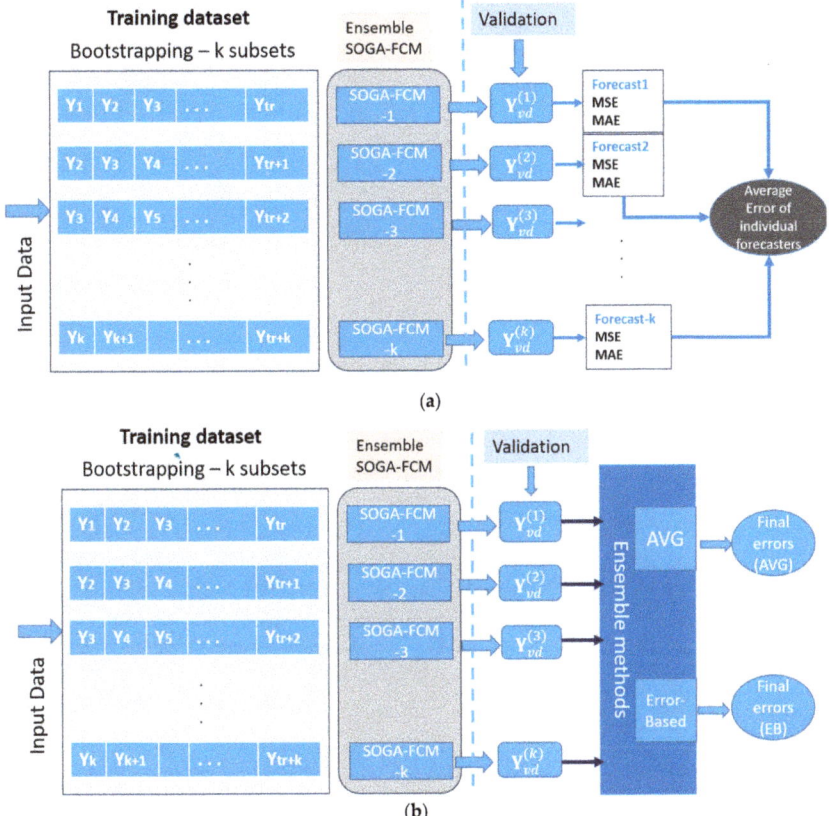

Figure 2. (a) Forecasting approach using individual forecasters of SOGA-FCM and mean average, (b) Example of the proposed forecasting combination approach for SOGA-FCM using ensemble methods. SOGA, structure optimization genetic algorithm; FCM, fuzzy cognitive map.

If we needed to accomplish daily forecasting, then we preselected the number of days excluded at each subset k. For the case of simplicity (as in the case of monthly forecasting), we could consider that one day was excluded at each sub-set from the initial in-sample dataset. The overall approach, including ANN, FCMs, and hybrid configurations of them, is illustrated in Figure 3. In Figure 3, the four ensemble forecasters were produced after the validation process and used for testing through the proposed approach.

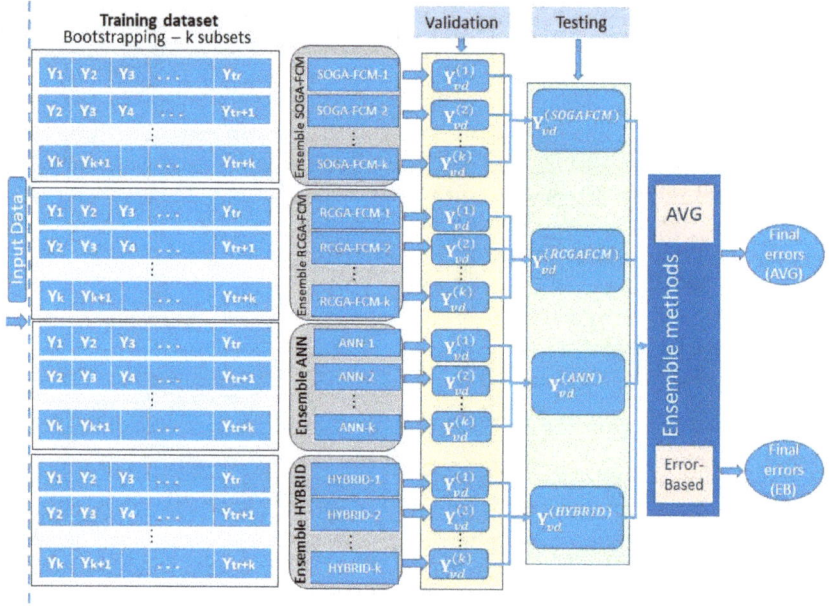

Figure 3. The proposed forecasting combination approach using ensemble methods and ensemble forecasters.

Step 3. We had n component forecasting models and obtained $\hat{Y}^i_{ts} = \left[\hat{y}^i_{N_{in}+1}, \hat{y}^i_{N_{in}+2}, \ldots, \hat{y}^i_{N_{in}+N_{ts}}\right]^T$ as the forecast of Y_{ts} through the ith model.

Step 4. We implemented each model on Y_{tr} and used it to predict Y_{vd}. Let $\hat{Y}^i_{vd} = \left[\hat{y}^i_{N_{tr}+1}, \hat{y}^i_{N_{tr}+2}, \ldots, \hat{y}^i_{N_{tr}+N_{vd}}\right]^T$ be the prediction of Y_{vd} through the ith model.

Step 5. We found the in-sample forecasting error of each model through some suitable error measures. We used the mean absolute error (MAE) and the mean squared error (MSE). These are widely popular error statistics [68], and their mathematical formulation is presented below in this paper. In the present study, we adopted the MSE and MAE to find the in-sample forecasting errors of the component models.

Step 6. Based on the obtained in-sample forecasting errors, we assigned a score to each component model as $\gamma_i = \frac{1}{MSE\ Y_{vd}, \hat{Y}^i_{vd}}$, $\forall i = 1, 2, \ldots, n$. The scores are assigned to be inversely proportional to the respective errors so that a model with a comparatively smaller in-sample error receives more score and vice versa.

Step 7. We assigned a rank $r_i \in 1, 2, \ldots, n$ to the ith model, based on its score, so that $r_i \geq r_j$, if $\gamma_i \leq \gamma_j$, $\forall i, j = 1, 2, \ldots, n$. The minimum, i.e., the best rank is equal to 1 and the maximum, i.e., the worst rank is at most equal to n.

Step 8. We chose a number n_r so that $1 \leq n_r \leq n$ and let $I = i_1, i_2, \ldots, i_{n_r}$ be the index set of the n_r component models, whose ranks are in the range $[1, n_r]$. So, we selected a subgroup of n_r smallest ranked component models.

Step 9. Finally, we obtained the weighted linear combination of these selected n_r component forecasts, as follows:

$$\hat{y}_k = w_{i_1}\hat{y}^{i_1}_k + w_{i_2}\hat{y}^{i_2}_k + \ldots + w_{i_{n_r}}\hat{y}^{i_{n_r}}_k = \sum_{i \in I} w_i \hat{y}^i_k, \quad \forall i = 1, 2, \ldots, n \qquad (19)$$

Here, $w_{i_k} = \frac{\gamma_{i_k}}{\sum_{k=1}^{n_r} \gamma_{i_k}}$ is the normalized weight to the selected component model, so that $\sum_{k=1}^{n_r} w_{i_k} = 1$.

Step 10. The simple average method could be also adopted, as an alternative to Step 6–9, to calculate the forecasted value.

The validation set was used during the training process for updating the algorithm weights appropriately and, thus, improving its performance and avoiding overfitting. After training the model, we could run it on the testing data, to verify if it has predicted them correctly and, if it has been so, to keep hidden the validation set.

The most popular and widely used performance metrics or evaluation criteria for time series prediction are the following: coefficient of determination (R2), mean square error (MSE), root mean square error (RMSE), mean absolute error (MAE), and mean absolute percentage error (MAPE). The mathematical equations of all these statistical indicators were described in the study [69]. The goodness of fit and the performance of the studied models, when they applied to a natural gas prediction process, were evaluated and compared using two of these five commonly used statistical indicators, namely, the MSE and the MAE [9]. In particular, the performance of the analyzed approaches for natural gas prediction was evaluated based on the following criteria:

1. Mean squared error:

$$\text{MSE} = \frac{1}{T} \sum_{t=1}^{T} (Z(t) - X(t))^2 \qquad (20)$$

2. Mean absolute error:

$$\text{MAE} = \frac{1}{T} \sum_{t=1}^{T} |Z(t) - X(t)| \qquad (21)$$

where $X(t)$ is the predicted value of the neural gas at the tth iteration, $Z(t)$ is the desired value of the neural gas at the tth iteration, $t = 1, \ldots, N_{ts}$, and N_{ts} is the number of testing records. The lower values of the MSE and MAE indicate that the model performance is better with respect to the prediction accuracy, and the regression line fits the data well.

All the modeling approaches, tests, and evaluations were performed with the use of the ISEMK (intelligent expert system based on cognitive maps) software tool [66], in which all the algorithms based on ANNs, FCMs, and their hybrid combinations were developed. C# programming language has been used for implementing ensemble models and also for developing ISEMK, which incorporates FCM construction from data and learning, both for RCGA and SOGA implementations [69].

4. Results and Discussion

4.1. Case Study and Datasets

The natural gas consumption datasets that were used in this research work to examine the applicability and effectiveness of the proposed forecast methodology corresponded to five years (2013–2017), as described in Section 3. Following the first step of the methodology, we split our dataset into training, validation, and testing ones. For the convenience of handling properly the dataset, we defined the data of the first three years as the training dataset (1095 days), the data of the fourth year as the validation dataset (365 days), and the remaining data (5th year) as the testing dataset (365 days), which approximately corresponded to 60%, 20%, and 20%, respectively, as presented in Section 3. Thus, it was easier for our analysis to handle the above values as annual datasets and have a clearer perception of the whole process.

Out of the three years of the defined training dataset, we used the first two as the initial training dataset, while the third (3rd) year was used as a dataset reservoir for the bootstrapping procedure. This year was properly selected to be part of the initial dataset, as for each value of k (the bootstrapping step), a corresponding number of days/weeks/months was additionally needed to be included in the training dataset during the bootstrapping process, thus, avoiding any possible data shortage and/or deterioration that would lead to inaccurate results.

The proposed forecast combination approach, presented in Section 3, offered generalization capabilities and, thus, it could be applied in various time-series datasets, for a different number of k, according to daily, weekly, or monthly prediction. Taking as an example the case of a month-ahead prediction, for each bootstrapping step k, the training dataset shifted one month ahead, getting one additional month each time from the reserved third year of the initial training dataset. In this case, k more months in total were needed for implementing efficiently this approach. If we considered $k = 12$, then 12 additional months of the initial dataset needed to be added and reserved. This approach justified our case where one year (i.e., the third year) was added to the initial training dataset and was further reserved for serving the purposes of the proposed methodology. Different values of k were also examined without noticing significant divergences in forecasting, compared to the selected k value.

In the next step, the validation procedure (comprising one year of data) was implemented to calculate the in-sample forecasting errors (MSE and MAE) for each ensemble forecasting algorithm (ensemble ANN, ensemble hybrid, ensemble RCGA-FCM, and ensemble SOGA-FCM). The same process was followed for the testing procedure by considering the data of the last year. The two examined ensemble forecasting methods, i.e., the simple average (AVG) and the error-based (EB), were then applied in the calculated validation and testing vectors (Yvd) for each one of the forecast combined methodology (Yvd-ANN, Yvd-Hybrid, Yvd-RCGA, Yvd-SOGA).

4.2. Case Study Results

In this study, we applied both the AVG and the EB method in two different cases: case (A) where scores were calculated for individual forecaster of each one of the methods ANN, hybrid, RCGA-FCM, and SOGA-FCM, and case (B), where scores were calculated for each ensemble forecaster (ANN ensemble, hybrid ensemble, RCGA-FCM ensemble, and SOGA-FCM ensemble).

Considering case (A), Table 1 shows the calculated errors and scores based on the EB method for individual forecaster of the two forecasting methods: ANN and hybrid for the city of Athens. The rest calculated errors and scores, based on the EB method, for individual forecaster for the other two remaining forecasting methods RCGA-FCM and SOGA-FCM for Athens can be found in Appendix A of the paper (Table A3). In Appendix A, parts of the corresponding results for the other two examined cities (Larissa and Thessaloniki) are also presented (Tables A4 and A5).

Table 1. Case (A)-Calculated errors and weights for each ensemble forecaster based on scores for EB (error-based) method (Athens).

	Validation		Testing				Testing		
	MAE	MSE	MAE	MSE	Weights		MAE	MSE	Weights
ANN1	0.0334	0.0035	0.0350	0.0036	0.2552	Hybrid1	0.0336	0.0034	0.2520
ANN2	0.0354	0.0041	0.0387	0.0043	0	Hybrid2	0.0387	0.0043	0
ANN3	0.0350	0.0037	0.0375	0.0039	0.2442	Hybrid3	0.0363	0.0037	0
ANN4	0.0341	0.0038	0.0365	0.0039	0	Hybrid4	0.0352	0.0035	0
ANN5	0.0335	0.0036	0.0358	0.0037	0.2505	Hybrid5	0.0339	0.0034	0
ANN6	0.0337	0.0039	0.0355	0.0038	0	Hybrid6	0.0348	0.0036	0.2468
ANN7	0.0336	0.0037	0.0362	0.0038	0	Hybrid7	0.0345	0.0035	0.2506
ANN8	0.0340	0.0039	0.0360	0.0039	0	Hybrid8	0.0354	0.0036	0
ANN9	0.0341	0.0039	0.0367	0.0040	0	Hybrid9	0.0349	0.0036	0
ANN10	0.0332	0.0036	0.0355	0.0037	0.2501	Hybrid10	0.0359	0.0038	0
ANN11	0.0338	0.0038	0.0365	0.0039	0	Hybrid11	0.0353	0.0038	0
ANN12	0.0345	0.0038	0.0349	0.0037	0	Hybrid12	0.0347	0.0033	0.2506
AVG	0.0336	0.0037	0.0359	0.0038		AVG	0.0350	0.0036	
EB	0.0335	0.0036	0.0358	0.0037		EB	0.0340	0.0034	

MSE: Mean Square Error, MAE: Mean Absolute Error.

Considering case (B), Table 2 presents the calculated weights based on scores for each ensemble forecaster (ANN (ensemble, hybrid ensemble, RCGA ensemble, and SOGA ensemble) for all three cities.

The calculated weights, based on scores for the EB method, were computed using Equation (17). According to this equation, the weights of the component forecasts are inversely proportional to their in-sample forecasting errors, concluding that the model with more error is assigned less weight to it and vice versa [80]. In this work, as the values of errors were high for certain ensemble forecasters, the corresponding weights were approximately zero, so they were considered to have a zero value for further predictions.

Table 2. Case (B)-Calculated weights for each ensemble forecaster based on scores for the EB method.

	Athens	Thessaloniki	Larissa
		Weights based on scores	
ANN	0.3320	0.34106	0.3369
Hybrid	0.3357	0.35162	0.3546
RCGA-FCM	0.3323	0	0
SOGA-FCM	0	0.30731	0.3083

ANN: Artificial Neural Network, RCGA-FCM: Real Codded Genetic Algorithm-Fuzzy Cognitive Map, SOGA-FCM: Structure Optimization Genetic Algorithm-Fuzzy Cognitive Map.

The obtained forecasting results of the individual and combination methods are depicted in Tables 3–8, respectively, for the three cities. In each of these tables, the best results (i.e., those associated with the least values of error measures) are presented in bold letters. In Figures 4 and 5, the forecasting results concerning Thessaloniki and Larissa are visually illustrated for both ensemble methods (AVG, EB). Moreover, Figure 6 gathers the forecasting results for all three cities considering the best ensemble method.

Table 3. Calculated errors for individual forecaster based on scores for Athens.

Validation	ANN	Hybrid	RCGA	SOGA	Ensemble AVG	Ensemble EB
MAE	0.0328	0.0333	0.0384	0.0391	0.0336	**0.0326**
MSE	0.0036	0.0035	0.0036	0.0037	**0.0032**	**0.0032**
Testing						
MAE	0.0321	0.0328	0.0418	0.0424	0.0345	0.0328
MSE	0.0033	0.0032	0.0038	0.0040	0.0032	**0.0031**

Table 4. Calculated errors for each ensemble forecaster based on scores for Athens.

Validation	ANN Ensemble	Hybrid Ensemble	RCGA Ensemble	SOGA Ensemble	Ensemble AVG	Ensemble EB
MAE	0.0335	0.0330	0.0388	0.0380	0.0337	0.0337
MSE	0.0036	0.0035	0.0036	0.0035	**0.0032**	**0.0032**
Testing						
MAE	0.0358	0.0340	0.0422	0.0422	0.0352	0.0352
MSE	0.0037	0.0034	0.0038	0.0037	**0.0032**	**0.0032**

Table 5. Calculated errors for individual forecaster based on scores for Thessaloniki.

Validation	ANN	Hybrid	RCGA	SOGA	Ensemble AVG	Ensemble EB
MAE	0.0343	0.0341	0.0381	0.0380	0.0347	0.0340
MSE	0.0029	0.0028	0.0032	0.0032	0.0028	**0.0027**
Testing						
MAE	0.0366	0.0381	0.0395	0.0399	0.0371	0.0369
MSE	0.0032	0.0033	0.0035	0.0036	0.0032	**0.0031**

Table 6. Calculated errors for each ensemble forecaster based on scores for Thessaloniki.

Validation	ANN Ensemble	Hybrid Ensemble	RCGA Ensemble	SOGA Ensemble	Ensemble AVG	Ensemble EB
MAE	0.0363	0.0361	0.0378	0.0374	0.0355	0.0355
MSE	0.0031	0.0031	0.0031	0.0030	**0.0028**	**0.0028**
Testing						
MAE	0.0393	0.0394	0.0399	0.0391	0.0381	0.0381
MSE	0.0037	0.0037	0.0036	0.0034	**0.0034**	**0.0034**

Table 7. Calculated errors for individual forecaster based on scores for Larissa.

Validation	ANN	Hybrid	RCGA	SOGA	Ensemble AVG	Ensemble EB
MAE	0.0322	0.0324	0.0372	0.0365	0.0326	0.0319
MSE	0.0030	0.0028	0.0033	0.0032	**0.0027**	**0.0027**
Testing						
MAE	0.0412	0.0417	0.0466	0.0468	0.0427	0.0417
MSE	0.0043	0.0041	0.0047	0.0047	**0.0040**	**0.0040**

Table 8. Case (B) -Calculated errors for each ensemble forecaster based on scores for Larissa.

Validation	ANN Ensemble	Hybrid Ensemble	RCGA Ensemble	SOGA Ensemble	Ensemble AVG	Ensemble EB
MAE	0.0337	0.0332	0.0371	0.0362	0.0329	0.0326
MSE	0.0032	0.0030	0.0032	0.0031	0.0027	**0.0026**
Testing						
MAE	0.0428	0.0417	0.0458	0.0460	0.0426	0.0423
MSE	0.0048	0.0044	0.0045	0.0045	0.0041	0.0040

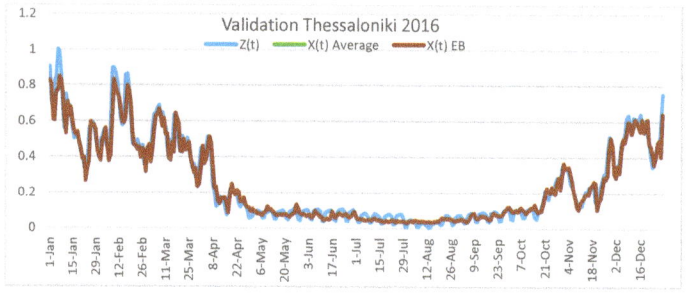

(a)

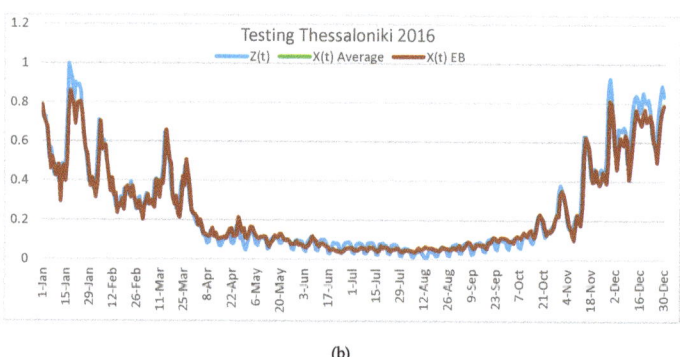

(b)

Figure 4. Forecasting results for Thessaloniki considering the two ensemble methods (AVG, EB) based on scores. (**a**) Validation, (**b**) Testing. AVG, simple average; EB, error-based.

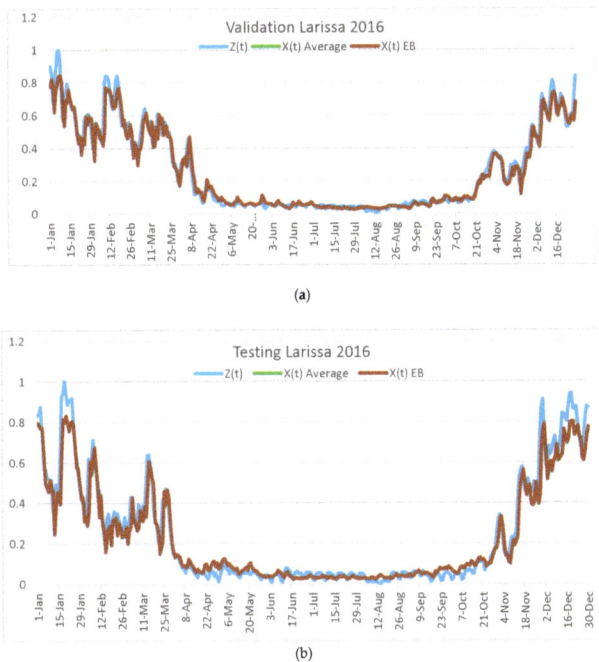

Figure 5. Forecasting results for Larissa considering the two ensemble methods (AVG, EB) based on scores. (**a**) Validation, (**b**) Testing.

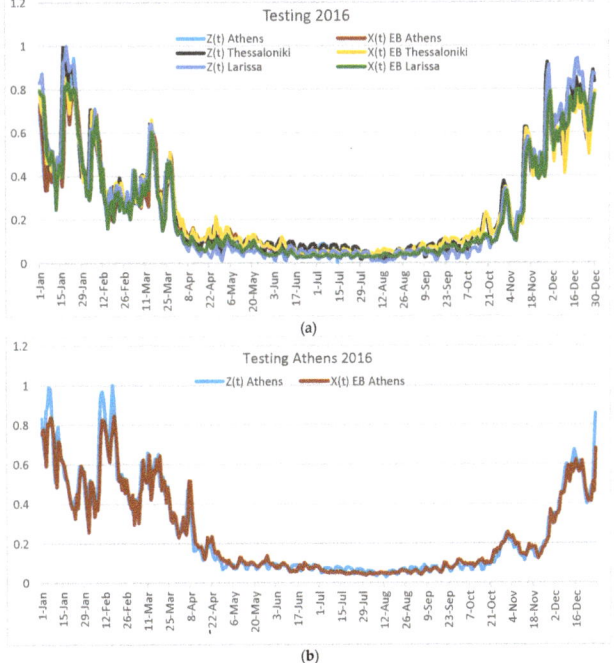

Figure 6. *Cont.*

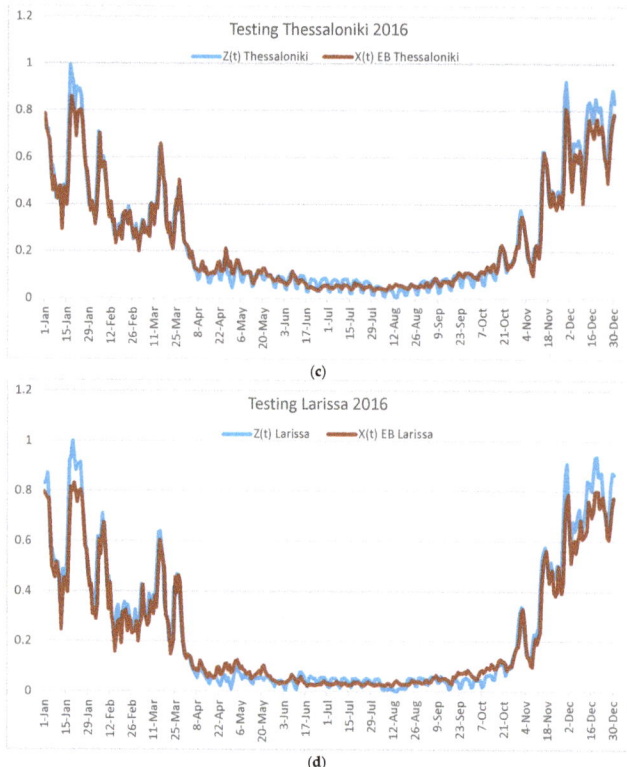

Figure 6. Forecasting results for the three cities considering the best ensemble method. (**a**) Testing all cities, (**b**) Testing Athens, (**c**) Testing Thessaloniki, (**d**) Testing Larissa.

4.3. Discussion of Results

The following important observations were noticed after careful analysis of Tables and Figures above.

1. After a thorough analysis of the Tables 3–8, on the basis of examining the MAE and MSE errors, it could be clearly stated that the EB method presented lower errors concerning the individual forecasters (ANN, hybrid, RCGA-FCM, and SOGA-FCM) for all three cities (Athens, Thessaloniki, and Larisa). EB seemed to outperform the AVG method in terms of achieving overall better forecasting results when applied to individual forecasters (see Figure 6).
2. Considering the ensemble forecasters, it could be seen from the obtained results that none of the two forecast combination methods had attained consistently better accuracies compared to each other, as far as the cities of Athens and Thessaloniki were concerned. Specifically, from Tables 3–6, it was observed that the MAE and MSE values across the two combination methods were similar for the two cities; however, their errors were lower than those produced by each separate ensemble forecaster.

3. Although the AVG and the EB methods performed similarly for Athens and Thessaloniki datasets, the EB forecast combination technique presented lower MAE and MSE errors than the AVG for the examined dataset of Larissa (see Figure 5).

The fact that when a forecasting method presented lower MAE and MSE errors than another means that the accuracy of the results produced with the first method, in terms of predicting consumption, was higher than the latter forecasting methods examined and compared to, so as the overall performance of the ensemble method was. Regarding the amount of improvement that was presented when a forecasting method was applied, slightly better performance of both ensemble forecasting methods could be noticed, and that constituted strong evidence for the efficiency of the examined method in the domain of natural gas demand forecasting.

In order to examine the efficiency of the proposed algorithm, a statistical test was conducted to reveal no statistical significance. Concerning the individual methods, a t-test paired two samples of mean was previously conducted in [60] for the cities of Thessaly (Larissa, Volos, Trikala, and Karditsa), for the year 2016, showing that there was no statistical significance among these techniques. In current work, a t-test paired two samples of mean was also performed, regarding the ensemble methods (average and error-based) for the examined cities (Athens, Thessaloniki, and Larissa), regarding the dataset of the same year. The results of the hypothesis tests (Tables A6–A8 in Appendix A) revealed no statistical significance between these techniques. In all cases, the calculated p-value exceeded 0.05, so no statistical significance was noticed from the obtained statistical analysis. Therefore, there was no particular need to conduct a post hoc statistical test, since a post hoc test should only be run when you have an overall statistically significant difference in group means, according to the relevant literature [89,90].

Furthermore, for comparison purposes, to show the effectiveness of the proposed forecasting combination approach of multivariate time series, the experimental analysis was conducted with a new and well-known effective machine learning technique for time series forecasting, the LSTM (long short-term memory). LSTM algorithm encloses the characteristics of the advanced recurrent neural network methods and is mainly applied for time series prediction problems in diverse domains [91].

LSTM was applied in one day-ahead natural gas consumption prediction concerning the same dataset of the three Greek cities (Athens, Thessaloniki, and Larissa) in [70]. For the LSTM implementation, one feature of the dataset as a time series was selected. As explained in [70], LSTM was fed previous values, and, in that case, the time-step was set to be 364 values to predict the next 364. For validation, 20% of random data from the training dataset was used, and for testing, the same dataset that was used for the ANN, RCGA-FCM, SOGA-FCM, and hybrid FCM-ANN, as well as with their ensemble structures implementation. In [70], various experiments with different numbers of units, number of layers, and dropout rates were accomplished. Through the provided experimental analysis, the best results of LSTM emerged for one layer, 200 units, and dropout rate = 0.2. These results are gathered in Table 9 for the three cities.

In Table 9, it is clear that both ensemble forecasting methods can achieve high accuracy in the predictions of the energy consumption patterns in a day-ahead timescale. Additional exploratory analysis and investigation of other types of ensemble methods, as well as other types of neural networks, such as convolutional neural networks (CNNs), could lead to a better insight of the modeling the particular problem and achieve higher prediction accuracy.

Table 9. Comparison results with LSTM (long short-term memory) (with best configuration parameters).

	Best Ensemble		LSTM (Dropout = 0.2)
	Case (A) (Individual)	Case (B) (Ensemble)	1 layer
Validation		ATHENS	
MAE	0.0326	0.0337	0.0406
MSE	0.0032	0.0032	0.0039
Testing			
MAE	0.0328	0.0352	0.0426
MSE	0.0031	0.0032	0.0041
Validation		THESSALONIKI	
MAE	0.0340	0.0355	0.0462
MSE	0.0027	0.0028	0.0043
Testing			
MAE	0.0369	0.0381	0.0489
MSE	0.0031	0.0034	0.0045
Validation		LARISSA	
MAE	0.0319	0.0326	0.0373
MSE	0.0027	0.0026	0.0029
Testing			
MAE	0.0417	0.0423	0.0462
MSE	0.0040	0.0040	0.0042

5. Conclusions

To sum up, we applied a time series forecasting method for natural gas demand in three Greek cities, implementing an efficient ensemble forecasting approach through combining ANN, RCGA-FCM, SOGA-FCM, and hybrid FCM-ANN. The proposed forecasting combination approach incorporates the two most popular ensemble methods for error calculation in forecasting problems and is deployed in certain steps offering generalization capabilities. The whole framework seems to be a promising approach for ensemble time series forecasting that can easily be applied in many scientific domains. An initial comparison analysis was conducted with benchmark methods of ANN, FCM, and their different configurations. Next, further comparison analysis was conducted with new promising LSTM networks previously used for time series prediction.

Through the experimental analysis, two error statistics (MAE, MSE) needed to be calculated in order to examine the effectiveness of the ensemble learning approach in time series prediction. The results of this study showed that the examined ensemble approach through designing an ensemble structure of various ANN, SOGA-FCM models by different learning parameters and their hybrid structures could significantly improve forecasting. Moreover, obtained results clearly demonstrated that a relatively higher forecasting accuracy was noticed when the applied ensemble approach was compared against independent forecasting approaches, such as ANN or FCM, as well as with LSTM.

Future work is devoted to applying the advantageous forecast combination approach to a larger number of distribution points that compose the natural gas grid of Greek regions (larger and smaller cities) as well as to investigate a new forecast combination structure of efficient convolutional neural networks (CNN) and LSTM networks for time series prediction in various application domains. Furthermore, an extensive comparative analysis with various LSTM structures, as well as with other advanced machine learning and time series prediction methods, will be conducted in future work. The presented research work could also contribute to explainability, transparency, and re-traceability of artificial intelligence (AI) and machine learning systems. These systems are being applied in various fields, and the decisions being made by them are not always clear due to the use of complicated algorithms in order to achieve power, performance, and accuracy. The authors with the use of complicated, but powerful algorithms, such as neural networks and ensemble methods, tried to describe all the steps and models involved in decision-making process to attain explainability and, in future, they would further explore ways to make the best-performing methods more transparent, re-traceable, and understandable, explaining why certain decisions have been made [92].

Author Contributions: For conceptualization, K.I.P.; methodology, K.I.P; software, K.P.; designed and performed the experiments, K.I.P. and K.P.; analyzed the results, drafted the initial manuscript, and revised the final manuscript, V.C.G., K.I.P., and E.P.; supervised the work, E.P., V.C.G., and G.S.

Funding: This research received no external funding.

Conflicts of Interest: The authors declare no conflict of interest.

Appendix A

Table A1. Descriptive statistics values for real dataset Z(t), forecasting values of AVG and EB methods for the three cities (validation).

Descriptive Statistics	Athens			Thessaloniki			Larissa		
	Z(t)	X(t)AVG	X(t) EB	Z(t)	X(t)AVG	X(t) EB	Z(t)	X(t)AVG	X(t) EB
Mean	0.2540	0.2464	0.2464	0.2611	0.2510	0.2510	0.2689	0.2565	0.2575
Median	0.1154	0.1366	0.1366	0.1335	0.1393	0.1394	0.1037	0.1194	0.1211
St. Deviation	0.2391	0.2203	0.2203	0.2373	0.2228	0.2228	0.2604	0.2429	0.2429
Kurtosis	0.3610	−0.2748	−0.2741	−0.1839	−0.5807	−0.5774	−0.6564	−0.8881	−0.8847
Skewness	1.1605	0.9801	0.9803	0.9328	0.8288	0.8298	0.8112	0.7520	0.7516
Minimum	0.0277	0.0367	0.0367	0.0043	0.0305	0.0304	0.0000	0.0235	0.0239
Maximum	1.0000	0.8429	0.8431	1.0000	0.8442	0.8448	1.0000	0.8361	0.8383

Table A2. Descriptive statistics values for real dataset Z(t), forecasting values of AVG and EB methods for the three cities (testing).

Descriptive Statistics	Athens			Thessaloniki			Larissa		
	Z(t)	X(t)AVG	X(t) EB	Z(t)	X(t)AVG	X(t) EB	Z(t)	X(t)AVG	X(t) EB
Mean	0.2479	0.2433	0.2433	0.2588	0.2478	0.2478	0.2456	0.2279	0.2291
Median	0.1225	0.1488	0.1488	0.1179	0.1304	0.1304	0.0695	0.0961	0.0972
St. Deviation	0.2159	0.2020	0.2021	0.2483	0.2236	0.2237	0.2742	0.2399	0.2404
Kurtosis	0.6658	0.2785	0.2792	0.1755	−0.1254	−0.1219	−0.0113	−0.1588	−0.1502
Skewness	1.2242	1.1138	1.1140	1.1348	1.0469	1.0479	1.1205	1.0900	1.0921
Minimum	0.0000	0.0359	0.0359	0.0079	0.0358	0.0357	0.0000	0.0233	0.0237
Maximum	0.9438	0.8144	0.8144	0.9950	0.8556	0.8562	1.0000	0.8291	0.8310

Table A3. Case (A)-Calculated errors and weights for each ensemble forecaster (RCGA and SOGA-FCM) based on scores for the EB method (Athens).

	Validation		Testing				Testing		
	MAE	MSE	MAE	MSE	Weights		MAE	MSE	Weights
RCGA1	0.0386	0.0036	0.0425	0.0038	0.2531	SOGA1	0.0435	0.0037	0.2520
RCGA2	0.0391	0.0038	0.0430	0.0039	0	SOGA2	0.0423	0.0038	0.2509
RCGA3	0.0399	0.0039	0.0428	0.0039	0	SOGA3	0.0425	0.0038	0
RCGA4	0.0384	0.0036	0.0419	0.0038	0.2522	SOGA4	0.0449	0.0042	0
RCGA5	0.0389	0.0037	0.0423	0.0039	0	SOGA5	0.0429	0.0040	0
RCGA6	0.0392	0.0036	0.0424	0.0039	0.2472	SOGA6	0.0432	0.0038	0.2494
RCGA7	0.0398	0.0038	0.0434	0.0041	0	SOGA7	0.0421	0.0039	0
RCGA8	0.0386	0.0037	0.0416	0.0039	0	SOGA8	0.0422	0.0039	0
RCGA9	0.0398	0.0036	0.0436	0.0041	0.2472	SOGA9	0.0434	0.0042	0
RCGA10	0.0388	0.0037	0.0417	0.0039	0	SOGA10	0.0422	0.0040	0
RCGA11	0.0393	0.0038	0.0419	0.0039	0	SOGA11	0.0420	0.0038	0.2475
RCGA12	0.0396	0.0037	0.0434	0.0041	0	SOGA12	0.0425	0.0040	0
AVG	0.0385	0.0036	0.0418	0.0038		AVG	0.0422	0.0039	
EB	0.0388	0.0036	0.0422	0.0038		EB	0.0422	0.0037	

Table A4. Case (A)-Calculated errors and weights for each ensemble forecaster based on scores for the EB method (Thessaloniki).

	Validation		Testing				Testing		
	MAE	MSE	MAE	MSE	Weights		MAE	MSE	Weights
Hybrid1	0.0356	0.0030	0.0390	0.0036	0.2565	SOGA1	0.0414	0.0040	0
Hybrid2	0.0381	0.0036	0.0409	0.0042	0	SOGA2	0.0417	0.0040	0
Hybrid3	0.0371	0.0032	0.0398	0.0039	0.2422	SOGA 3	0.0394	0.0034	0
Hybrid4	0.0376	0.0032	0.0403	0.0039	0	SOGA 4	0.0406	0.0038	0
Hybrid5	0.0373	0.0032	0.0401	0.0040	0	SOGA 5	0.0388	0.0033	0.2541
Hybrid6	0.0375	0.0033	0.0403	0.0040	0	SOGA 6	0.0413	0.0038	0
Hybrid7	0.0378	0.0033	0.0405	0.0040	0	SOGA 7	0.0415	0.0039	0
Hybrid8	0.0373	0.0032	0.0402	0.0040	0	SOGA 8	0.0399	0.0036	0
Hybrid9	0.0378	0.0034	0.0407	0.0041	0	SOGA 9	0.0392	0.0035	0.2448
Hybrid10	0.0371	0.0032	0.0397	0.0039	0.2410	SOGA10	0.0400	0.0037	0
Hybrid11	0.0370	0.0033	0.0402	0.0040	0	SOGA11	0.0403	0.0036	0.2439
Hybrid12	0.0364	0.0030	0.0406	0.0036	0.2601	SOGA12	0.0397	0.0034	0.2569
AVG	0.0369	0.0032	0.0398	0.0039		AVG	0.0398	0.0036	
EB	0.0361	0.0031	0.0394	0.0037		EB	0.0391	0.0034	

Table A5. Case (A)-Calculated errors and weights for each ensemble forecaster based on scores for the EB method (Larissa).

	Validation		Testing				Testing		
	MAE	MSE	MAE	MSE	Weights		MAE	MSE	Weights
ANN1	0.0339	0.0032	0.0425	0.0047	0.2511	Hybrid1	0.0411	0.0043	0.2531
ANN2	0.0353	0.0036	0.0438	0.0052	0	Hybrid2	0.0435	0.0051	0
ANN3	0.0343	0.0033	0.0433	0.0050	0	Hybrid3	0.0418	0.0045	0.2472
ANN4	0.0347	0.0033	0.0429	0.0049	0	Hybrid4	0.0424	0.0048	0
ANN5	0.0353	0.0035	0.0436	0.0051	0	Hybrid5	0.0436	0.0051	0
ANN6	0.0352	0.0035	0.0432	0.0049	0	Hybrid6	0.0436	0.0051	0
ANN7	0.0354	0.0035	0.0441	0.0053	0	Hybrid7	0.0434	0.0050	0
ANN8	0.0348	0.0033	0.0427	0.0049	0	Hybrid8	0.0425	0.0047	0.2398
ANN9	0.0351	0.0035	0.0439	0.0052	0	Hybrid9	0.0423	0.0047	0
ANN10	0.0343	0.0033	0.0431	0.0049	0.2406	Hybrid10	0.0432	0.0050	0
ANN11	0.0342	0.0032	0.0436	0.0049	0.2472	Hybrid11	0.0444	0.0053	0
ANN12	0.0331	0.0031	0.0428	0.0047	0.2610	Hybrid12	0.0426	0.0043	0.2597
AVG	0.0345	0.0033	0.0431	0.0049		AVG	0.0427	0.0048	
EB	0.0337	0.0032	0.0428	0.0048		EB	0.0417	0.0044	

Table A6. t-Test: Paired Two Sample for Means between the ensemble methods (AVG and EB) (Athens).

	X(t) AVG Athens	X(t) EB Athens
Mean	0.243342155	0.243346733
Variance	0.040822427	0.040826581
Observations	196	196
Pearson Correlation	0.99999997	
Hypothesized Mean Difference	0	
df	195	
t Stat	−1.278099814	
P(T<=t) one-tail	0.101366761	
t Critical one-tail	1.65270531	
P(T<=t) two-tail	0.202733521	
t Critical two-tail	1.972204051	

Table A7. t-Test: Paired Two Sample for Means between the ensemble methods (AVG and EB) (Thessaloniki).

	X(t) AVG	X(t) EB
Mean	0.247811356	0.247811056
Variance	0.050004776	0.050032786
Observations	365	365
Pearson Correlation	0.999999788	
Hypothesized Mean Difference	0	
df	364	
t Stat	0.036242052	
P(T<=t) one-tail	0.485554611	
t Critical one-tail	1.649050545	
P(T<=t) two-tail	0.971109222	
t Critical two-tail	1.966502569	

Table A8. t-Test: Paired Two Sample for Means between the ensemble methods (AVG and EB) (Larissa).

	X(t) AVG Larisa	X(t) EB Larisa
Mean	0.227903242	0.229120614
Variance	0.057542802	0.05781177
Observations	365	365
Pearson Correlation	0.999972455	
Hypothesized Mean Difference	0	
df	364	
t Stat	−12.44788062	
P(T<=t) one-tail	3.52722E-30	
t Critical one-tail	1.649050545	
P(T<=t) two-tail	7.05444E-30	
t Critical two-tail	1.966502569	

References

1. Li, H.Z.; Guo, S.; Li, C.J.; Sun, J.Q. A hybrid annual power load forecasting model based on generalized regression neural network with fruit fly optimization algorithm. *Knowl.-Based Syst.* **2013**, *37*, 378–387. [CrossRef]
2. Bodyanskiy, Y.; Popov, S. Neural network approach to forecasting of quasiperiodic financial time series. *Eur. J. Oper. Res.* **2006**, *175*, 1357–1366. [CrossRef]
3. Livieris, I.E.; Kotsilieris, T.; Stavroyiannis, S.; Pintelas, P. Forecasting stock price index movement using a constrained deep neural network training algorithm. *Intell. Decis. Technol.* **2019**, 1–14. Available online: https://www.researchgate.net/publication/334132665_Forecasting_stock_price_index_movement_using_a_constrained_deep_neural_network_training_algorithm (accessed on 10 September 2019).
4. Livieris, I.E.; Pintelas, H.; Kotsilieris, T.; Stavroyiannis, S.; Pintelas, P. Weight-constrained neural networks in forecasting tourist volumes: A case study. *Electronics* **2019**, *8*, 1005. [CrossRef]
5. Chen, C.F.; Lai, M.C.; Yeh, C.C. Forecasting tourism demand based on empirical mode decomposition and neural network. *Knowl.-Based Syst.* **2012**, *26*, 281–287. [CrossRef]
6. Lu, X.; Wang, J.; Cai, Y.; Zhao, J. Distributed HS-ARTMAP and its forecasting model for electricity load. *Appl. Soft Comput.* **2015**, *32*, 13–22. [CrossRef]
7. Zeng, Y.R.; Zeng, Y.; Choi, B.; Wang, L. Multifactor-influenced energy consumption forecasting using enhanced back-propagation neural network. *Energy* **2017**, *127*, 381–396. [CrossRef]
8. Lin, W.Y.; Hu, Y.H.; Tsai, C.F. Machine learning in financial crisis prediction: A survey. *IEEE Trans. Syst. Man Cybern.* **2012**, *42*, 421–436.
9. Raza, M.Q.; Khosravi, A. A review on artificial intelligence-based load demand techniques for smart grid and buildings. *Renew. Sustain. Energy Rev.* **2015**, *50*, 1352–1372. [CrossRef]
10. Weron, R. Electricity price forecasting: A review of the state-of-the-art with a look into the future. *Int. J. Forecast.* **2014**, *30*, 1030–1081. [CrossRef]

11. Meade, N.; Islam, T. Forecasting in telecommunications and ICT-A review. *Int. J. Forecast.* **2015**, *31*, 1105–1126. [CrossRef]
12. Donkor, E.; Mazzuchi, T.; Soyer, R.; Alan Roberson, J. Urban water demand forecasting: Review of methods and models. *J. Water Resour. Plann. Managem.* **2014**, *140*, 146–159. [CrossRef]
13. Fagiani, M.; Squartini, S.; Gabrielli, L.; Spinsante, S.; Piazza, F. A review of datasets and load forecasting techniques for smart natural gas and water grids: Analysis and experiments. *Neurocomputing* **2015**, *170*, 448–465. [CrossRef]
14. Chandra, D.R.; Kumari, M.S.; Sydulu, M. A detailed literature review on wind forecasting. In Proceedings of the IEEE International Conference on Power, Energy and Control (ICPEC), Sri Rangalatchum Dindigul, India, 6–8 February 2013; pp. 630–634.
15. Zhang, G. Time series forecasting using a hybrid ARIMA and neural network model. *Neurocomputing* **2003**, *50*, 159–175. [CrossRef]
16. Adhikari, R.; Verma, G.; Khandelwal, I. A model ranking based selective ensemble approach for time series forecasting. *Procedia Comput. Sci.* **2015**, *48*, 14–21. [CrossRef]
17. Box, G.E.P.; Jenkins, G.M.; Reinsel, G.C. *Time Series Analysis: Forecasting and Control*, 3rd ed.; Prentice-Hall: Englewood Cliffs, NJ, USA, 1994.
18. Kumar, K.; Jain, V.K. Autoregressive integrated moving averages (ARIMA) modeling of a traffic noise time series. *Appl. Acoust.* **1999**, *58*, 283–294. [CrossRef]
19. Ediger, V.S.; Akar, S. ARIMA forecasting of primary energy demand by fuel in Turkey. *Energy Policy* **2007**, *35*, 1701–1708. [CrossRef]
20. Poczęta, K.; Kubus, L.; Yastrebov, A.; Papageorgiou, E.I. Temperature forecasting for energy saving in smart buildings based on fuzzy cognitive map. In Proceedings of the 2018 Conference on Automation (Automation 2018), Warsaw, Poland, 21–23 March 2018; pp. 93–103.
21. Poczęta, K.; Yastrebov, A.; Papageorgiou, E.I. Application of fuzzy cognitive maps to multi-step ahead prediction of electricity consumption. In Proceedings of the IEEE 2018 Conference on Electrotechnology: Processes, Models, Control and Computer Science (EPMCCS), Kielce, Poland, 12–14 November 2018; pp. 1–5.
22. Khashei, M.; Rafiei, F.M.; Bijari, M. Hybrid fuzzy auto-regressive integrated moving average (FARIMAH) model for forecasting the foreign exchange markets. *Int. J. Comput. Intell. Syst.* **2013**, *6*, 954–968. [CrossRef]
23. Chu, F.L. A fractionally integrated autoregressive moving average approach to forecasting tourism demand. *Tour. Manag.* **2008**, *29*, 79–88. [CrossRef]
24. Yu, H.K.; Kim, N.Y.; Kim, S.S. Forecasting the number of human immunodeficiency virus infections in the Korean population using the autoregressive integrated moving average model. *Osong Public Health Res. Perspect.* **2013**, *4*, 358–362. [CrossRef]
25. Haykin, S. *Neural Networks: A Comprehensive Foundation*, 2nd ed.; Prentice Hall: Upper Saddle River, NJ, USA, 1999.
26. Doganis, P.; Alexandridis, A.; Patrinos, P.; Sarimveis, H. Time series sales forecasting for short shelf-life food products based on artificial neural networks and evolutionary computing. *J. Food Eng.* **2006**, *75*, 196–204. [CrossRef]
27. Kim, M.; Jeong, J.; Bae, S. Demand forecasting based on machine learning for mass customization in smart manufacturing. In Proceedings of the ACM 2019 International Conference on Data Mining and Machine Learning (ICDMML 2019), Hong Kong, China, 28–30 April 2019; pp. 6–11.
28. Wang, D.M.; Wang, L.; Zhang, G.M. Short-term wind speed forecast model for wind farms based on genetic BP neural network. *J. Zhejiang Univ. (Eng. Sci.)* **2012**, *46*, 837–842.
29. Breiman, L. Stacked regressions. *Mach. Learn.* **1996**, *24*, 49–64. [CrossRef]
30. Clemen, R. Combining forecasts: A review and annotated bibliography. *J. Forecast.* **1989**, *5*, 559–583. [CrossRef]
31. Perrone, M. Improving Regression Estimation: Averaging Methods for Variance Reduction with Extension to General Convex Measure Optimization. Ph.D. Thesis, Brown University, Providence, RI, USA, 1993.
32. Wolpert, D. Stacked generalization. *Neural Netw.* **1992**, *5*, 241–259. [CrossRef]
33. Geman, S.; Bienenstock, E.; Doursat, R. Neural networks and the bias/variance dilemma. *Neural Comput.* **1992**, *4*, 1–58. [CrossRef]
34. Krogh, A.; Sollich, P. Statistical mechanics of ensemble learning. *Phys. Rev.* **1997**, *55*, 811–825. [CrossRef]

35. Freund, Y.; Schapire, R. Experiments with a new boosting algorithm. In Proceedings of the 13th International Conference on Machine Learning (ICML 96), Bari, Italy, 3–6 July 1996; pp. 148–156.
36. Schapire, R. The strength of weak learnability. *Mach. Learn.* **1990**, *5*, 197–227. [CrossRef]
37. Che, J. Optimal sub-models selection algorithm for combination forecasting model. *Neurocomputing* **2014**, *151*, 364–375. [CrossRef]
38. Wei, N.; Li, C.J.; Li, C.; Xie, H.Y.; Du, Z.W.; Zhang, Q.S.; Zeng, F.H. Short-term forecasting of natural gas consumption using factor selection algorithm and optimized support vector regression. *J. Energy Resour. Technol.* **2018**, *141*, 032701. [CrossRef]
39. Tamba, J.G.; Essiane, S.N.; Sapnken, E.F.; Koffi, F.D.; Nsouandélé, J.L.; Soldo, B.; Njomo, D. Forecasting natural gas: A literature survey. *Int. J. Energy Econ. Policy* **2018**, *8*, 216–249.
40. Sebalj, D.; Mesaric, J.; Dujak, D. Predicting natural gas consumption a literature review. In Proceedings of the 28th Central European Conference on Information and Intelligent Systems (CECIIS '17), Varazdin, Croatia, 27–29 September 2017; pp. 293–300.
41. Azadeh, I.P.; Dagoumas, A.S. Day-ahead natural gas demand forecasting based on the combination of wavelet transform and ANFIS/genetic algorithm/neural network model. *Energy* **2017**, *118*, 231–245.
42. Gorucu, F.B.; Geris, P.U.; Gumrah, F. Artificial neural network modeling for forecasting gas consumption. *Energy Sources* **2004**, *26*, 299–307. [CrossRef]
43. Karimi, H.; Dastranj, J. Artificial neural network-based genetic algorithm to predict natural gas consumption. *Energy Syst.* **2014**, *5*, 571–581. [CrossRef]
44. Khotanzad, A.; Erlagar, H. Natural gas load forecasting with combination of adaptive neural networks. In Proceedings of the International Joint Conference on Neural Networks (IJCNN 99), Washington, DC, USA, 10–16 July 1999; pp. 4069–4072.
45. Khotanzad, A.; Elragal, H.; Lu, T.L. Combination of artificial neural-network forecasters for prediction of natural gas consumption. *IEEE Trans. Neural Netw.* **2000**, *11*, 464–473. [CrossRef]
46. Kizilaslan, R.; Karlik, B. Combination of neural networks forecasters for monthly natural gas consumption prediction. *Neural Netw. World* **2009**, *19*, 191–199.
47. Kizilaslan, R.; Karlik, B. Comparison neural networks models for short term forecasting of natural gas consumption in Istanbul. In Proceedings of the First International Conference on the Applications of Digital Information and Web Technologies (ICADIWT), Ostrava, Czech Republic, 4–6 August 2008; pp. 448–453.
48. Musilek, P.; Pelikan, E.; Brabec, T.; Simunek, M. Recurrent neural network based gating for natural gas load prediction system. In Proceedings of the 2006 IEEE International Joint Conference on Neural Networks, Vancouver, BC, Canada, 16–21 July 2006; pp. 3736–3741.
49. Soldo, B. Forecasting natural gas consumption. *Appl. Energy* **2012**, *92*, 26–37. [CrossRef]
50. Szoplik, J. Forecasting of natural gas consumption with artificial neural networks. *Energy* **2015**, *85*, 208–220. [CrossRef]
51. Potočnik, P.; Soldo, B.; Šimunović, G.; Šarić, T.; Jeromen, A.; Govekar, E. Comparison of static and adaptive models for short-term residential natural gas forecasting in Croatia. *Appl. Energy* **2014**, *129*, 94–103. [CrossRef]
52. Akpinar, M.; Yumusak, N. Forecasting household natural gas consumption with ARIMA model: A case study of removing cycle. In Proceedings of the 7th IEEE International Conference on Application of Information and Communication Technologies (ICAICT 2013), Baku, Azerbaijan, 23–25 October 2013; pp. 1–6.
53. Akpinar, M.M.; Adak, F.; Yumusak, N. Day-ahead natural gas demand forecasting using optimized ABC-based neural network with sliding window technique: The case study of regional basis in Turkey. *Energies* **2017**, *10*, 781. [CrossRef]
54. Taşpınar, F.; Çelebi, N.; Tutkun, N. Forecasting of daily natural gas consumption on regional basis in Turkey using various computational methods. *Energy Build.* **2013**, *56*, 23–31. [CrossRef]
55. Azadeh, A.; Asadzadeh, S.M.; Ghanbari, A. An adaptive network-based fuzzy inference system for short-term natural gas demand estimation: Uncertain and complex environments. *Energy Policy* **2010**, *38*, 1529–1536. [CrossRef]
56. Behrouznia, A.; Saberi, M.; Azadeh, A.; Asadzadeh, S.M.; Pazhoheshfar, P. An adaptive network based fuzzy inference system-fuzzy data envelopment analysis for gas consumption forecasting and analysis: The case of South America. In Proceedings of the 2010 International Conference on Intelligent and Advanced Systems (ICIAS), Manila, Philippines, 15–17 June 2010; pp. 1–6.

57. Viet, N.H.; Mandziuk, J. Neural and fuzzy neural networks for natural gas prediction consumption. In Proceedings of the 13th IEEE Workshop on Neural Networks for Signal Processing (NNSP 2003), Toulouse, France, 17–19 September 2003; pp. 759–768.
58. Yu, F.; Xu, X. A short-term load forecasting model of natural gas based on optimized genetic algorithm and improved BP neural network. *Appl. Energy* **2014**, *134*, 102–113. [CrossRef]
59. Papageorgiou, E.I. *Fuzzy Cognitive Maps for Applied Sciences and Engineering from Fundamentals to Extensions and Learning Algorithms*; Intelligent Systems Reference Library, Springer: Heidelberg, Germany, 2014.
60. Poczeta, K.; Papageorgiou, E.I. Implementing fuzzy cognitive maps with neural networks for natural gas prediction. In Proceedings of the 30th IEEE International Conference on Tools with Artificial Intelligence (ICTAI 2018), Volos, Greece, 5–7 November 2018; pp. 1026–1032.
61. Homenda, W.; Jastrzebska, A.; Pedrycz, W. Modeling time series with fuzzy cognitive maps. In Proceedings of the 2014 IEEE International Conference on Fuzzy Systems (FUZZ-IEEE), Beijing, China, 6–11 July 2014; pp. 2055–2062.
62. Homenda, W.; Jastrzębska, A.; Pedrycz, W. Nodes selection criteria for fuzzy cognitive maps designed to model time series. *Adv. Intell. Syst. Comput.* **2015**, *323*, 859–870.
63. Salmeron, J.L.; Froelich, W. Dynamic optimization of fuzzy cognitive maps for time series forecasting. *Knowl -Based Syst.* **2016**, *105*, 29–37. [CrossRef]
64. Froelich, W.; Salmeron, J.L. Evolutionary learning of fuzzy grey cognitive maps for the forecasting of multivariate, interval-valued time series. *Int. J. Approx. Reason* **2014**, *55*, 1319–1335. [CrossRef]
65. Papageorgiou, E.I.; Poczęta, K.; Laspidou, C. Application of fuzzy cognitive maps to water demand prediction, fuzzy systems. In Proceedings of the 2015 IEEE International Conference on Fuzzy Systems (FUZZ-IEEE), Istanbul, Turkey, 2–5 August 2015; pp. 1–8.
66. Poczeta, K.; Yastrebov, A.; Papageorgiou, E.I. Learning fuzzy cognitive maps using structure optimization genetic algorithm. In Proceedings of the 2015 Federated Conference on Computer Science and Information Systems (FedCSIS), Lodz, Poland, 13–16 September 2015; pp. 547–554.
67. Jastriebow, A.; Poczęta, K. Analysis of multi-step algorithms for cognitive maps learning. *Bull. Pol. Acad. Sci. Technol. Sci.* **2014**, *62*, 735–741. [CrossRef]
68. Stach, W.; Kurgan, L.; Pedrycz, W.; Reformat, M. Genetic learning of fuzzy cognitive maps. *Fuzzy Sets Syst.* **2005**, *153*, 371–401. [CrossRef]
69. Papageorgiou, E.I.; Poczeta, K. A two-stage model for time series prediction based on fuzzy cognitive maps and neural networks. *Neurocomputing* **2017**, *232*, 113–121. [CrossRef]
70. Anagnostis, A.; Papageorgiou, E.; Dafopoulos, V.; Bochtis, D. Applying Long Short-Term Memory Networks for natural gas demand prediction. In Proceedings of the 10th International Conference on Information, Intelligence, Systems and Applications (IISA 2019), Patras, Greece, 15–17 July 2019.
71. Kosko, B. Fuzzy cognitive maps. *Int. J. Man Mach. Stud.* **1986**, *24*, 65–75. [CrossRef]
72. Kardaras, D.; Mentzas, G. Using fuzzy cognitive maps to model and analyze business performance assessment. In *Advances in Industrial Engineering Applications and Practice II*; Chen, J., Mital, A., Eds.; International Journal of Industrial Engineering: San Diego, CA, USA, 12–15 November 1997; pp. 63–68.
73. Lee, K.Y.; Yang, F.F. Optimal reactive power planning using evolutionary algorithms: A comparative study for evolutionary programming, evolutionary strategy, genetic algorithm, and linear programming. *IEEE Trans. Power Syst.* **1998**, *13*, 101–108. [CrossRef]
74. Dickerson, J.A.; Kosko, B. Virtual Worlds as Fuzzy Cognitive Maps. *Presence* **1994**, *3*, 173–189. [CrossRef]
75. Hossain, S.; Brooks, L. Fuzzy cognitive map modelling educational software adoption. *Comput. Educ.* **2008**, *51*, 1569–1588. [CrossRef]
76. Fogel, D.B. *Evolutionary Computation. Toward a New Philosophy of Machine Intelligence*, 3rd ed.; Wiley: Hoboken, NJ, USA; IEEE Press: Piscataway, NJ, USA, 2006.
77. Herrera, F.; Lozano, M.; Verdegay, J.L. Tackling real-coded genetic algorithms: Operators and tools for behavioural analysis. *Artif. Intell. Rev.* **1998**, *12*, 265–319. [CrossRef]
78. Livieris, I.E. Forecasting Economy-related data utilizing weight-constrained recurrent neural networks. *Algorithms* **2019**, *12*, 85. [CrossRef]
79. Livieris, I.E.; Pintelas, P. *A Survey on Algorithms for Training Artificial Neural Networks*; Technical Report; Department of Mathematics, University of Patras: Patras, Greece, 2008.

80. Adhikari, R.; Agrawal, R.K. Performance evaluation of weight selection schemes for linear combination of multiple forecasts. *Artif. Intell. Rev.* **2012**, *42*, 1–20. [CrossRef]
81. Livieris, I.E.; Kanavos, A.; Tampakas, V.; Pintelas, P. A weighted voting ensemble self-labeled algorithm for the detection of lung abnormalities from X-rays. *Algorithms* **2019**, *12*, 64. [CrossRef]
82. Makridakis, S.; Winkler, R.L. Averages of forecasts: Some empirical results. *Manag. Sci.* **1983**, *29*, 987–996. [CrossRef]
83. Lemke, C.; Gabrys, B. Meta-learning for time series forecasting and forecast combination. *Neurocomputing* **2010**, *73*, 2006–2016. [CrossRef]
84. Palm, F.C.; Zellner, A. To combine or not to combine? Issues of combining forecasts. *J. Forecast.* **1992**, *11*, 687–701. [CrossRef]
85. Kreiss, J.; Lahiri, S. Bootstrap methods for time series. In *Time Series Analysis: Methods and Applications*; Rao, T., Rao, S., Rao, C., Eds.; Elsevier: Amsterdam, The Netherlands, 2012.
86. Shalabi, L.A.; Shaaban, Z. Normalization as a preprocessing engine for data mining and the approach of preference matrix. In Proceedings of the International Conference on Dependability of Computer Systems (DepCos-RELCOMEX), Szklarska Poreba, Poland, 25–27 May 2006; pp. 207–214.
87. Nayak, S.; Misra, B.B.; Behera, H. Evaluation of normalization methods on neuro-genetic models for stock index forecasting. In Proceedings of the 2012 World Congress on Information and Communication Technologies (WICT 2012), Trivandrum, India, 30 October–2 November 2012; pp. 602–607.
88. Jayalakshmi, T.; Santhakumaran, A. Statistical normalization and back propagation for classification. *Int. J. Comput. Theory Eng.* **2011**, *3*, 89–93. [CrossRef]
89. Brown, A. A new software for carrying out one-way ANOVA post hoc tests. *Comput. Methods Programs Biomed.* **2005**, *79*, 89–95. [CrossRef] [PubMed]
90. Chen, T.; Xu, M.; Tu, J.; Wang, H.; Niu, X. Relationship between omnibus and post-hoc tests: An investigation of performance of the F test in ANOVA. *Shanghai Arch. Psychiatry* **2018**, *30*, 60–64. [PubMed]
91. Understanding the LSTM Networks. Available online: http://colah.github.io (accessed on 10 September 2019).
92. Holzinger, A. From machine learning to explainable AI. In Proceedings of the 2018 World Symposium on Digital Intelligence for Systems and Machines (IEEE DISA), Kosice, Slovakia, 23–25 August 2018; pp. 55–66.

© 2019 by the authors. Licensee MDPI, Basel, Switzerland. This article is an open access article distributed under the terms and conditions of the Creative Commons Attribution (CC BY) license (http://creativecommons.org/licenses/by/4.0/).

Article

A Grey-Box Ensemble Model Exploiting Black-Box Accuracy and White-Box Intrinsic Interpretability

Emmanuel Pintelas [1], Ioannis E. Livieris [2,*] and Panagiotis Pintelas [2]

1 Department of Electrical & Computer Engineering, University of Patras, GR 265-00 Patras, Greece; manolis.pintelas@hotmail.com
2 Department of Mathematics, University of Patras, GR 265-00 Patras, Greece; ppintelas@gmail.com
* Correspondence: livieris@upatras.gr

Received: 14 November 2019; Accepted: 2 January 2020; Published: 5 January 2020

Abstract: Machine learning has emerged as a key factor in many technological and scientific advances and applications. Much research has been devoted to developing high performance machine learning models, which are able to make very accurate predictions and decisions on a wide range of applications. Nevertheless, we still seek to understand and explain how these models work and make decisions. Explainability and interpretability in machine learning is a significant issue, since in most of real-world problems it is considered essential to understand and explain the model's prediction mechanism in order to trust it and make decisions on critical issues. In this study, we developed a Grey-Box model based on semi-supervised methodology utilizing a self-training framework. The main objective of this work is the development of a both interpretable and accurate machine learning model, although this is a complex and challenging task. The proposed model was evaluated on a variety of real world datasets from the crucial application domains of education, finance and medicine. Our results demonstrate the efficiency of the proposed model performing comparable to a Black-Box and considerably outperforming single White-Box models, while at the same time remains as interpretable as a White-Box model.

Keywords: explainable machine learning; interpretable machine learning; semi-supervised learning; self-training algorithms; ensemble learning; black, white and grey box models

1. Introduction

Machine learning (ML) is a growing and popular field which has been successfully applied in many application domains. In particular, ML is a subfield of artificial intelligence which utilizes algorithmic models to extract useful and valuable information from large amounts of data in order to build intelligent computational systems which make decisions on specific problems in various fields such as finance, banking, medicine and education. However, in most of real-world applications, finding sufficient labeled data is a hard task while in contrast finding unlabeled data is an easier task [1,2]. To address this problem, Semi-supervised learning (SSL) possesses a proper methodology to exploit the hidden information in an unlabeled dataset aiming to build more efficient classifiers. It is based on the philosophy that both labeled and unlabeled data can be used to build efficient and accurate ML models [3].

Self-labeled algorithms form the most widely utilized and successful class of SSL algorithms which attempt to enlarge an initial small labeled dataset from a rather large unlabeled dataset [4]. The labeling of the unlabeled instances is achieved via an iterative labeling process of the algorithm's most confident predictions. One of the most popular and efficient self-labeled methods proposed in the literature is self-training. This algorithm initially trains one classifier on a labeled dataset in order to make predictions on an unlabeled dataset. Then, following an iterative procedure, the most confident predictions are added to the labeled set and the classifier is retrained on the new augmented dataset. When the iterations stop, this classifier is used for the final predictions on the test set.

In general, a classification model must be accurate in its predictions. However, another significant factor in ML models is the explainability and the interpretability of classifiers' predictions [5–7]. There are many real-world tasks which place significant importance in the understanding and the explanation of the model's prediction [8], especially in crucial and decisive decisions such as cancer diagnosis. Thus, research must focus on finding ML models which are both accurate and interpretable, although this is a complex and challenging task. The problem is that most of ML models which are interpretable (White-Boxes in software engineering terms) are less accurate, while models which are very accurate are hard or impossible to interpret (Black-Boxes). Therefore, a Grey-Box model which is a combination of black and White-Box models, may acquire the benefits of both, leading to an efficient ML model which could be both accurate and interpretable at the same time.

The innovation and the purpose of this work lies in the development of a Grey-Box ML model, based on the self-training framework, which is nearly as accurate as a Black-Box and more accurate than a single White-Box but is also interpretable like a White-Box model. More specifically, we utilized a Black-Box model in order to enlarge a small initial labeled dataset adding the model's most confident predictions of a large unlabeled dataset. Then, this augmented dataset is utilized as a training set on a White-Box model. By this combination, we aim to obtain the benefits of both models, which are Black-Box's high classification accuracy and White-Box's interpretability, in order to build a both accurate and interpretable ensemble.

The remainder of this paper is organized as follows: Section 2 presents in a more detailed way the meaning and the importance of interpretability and explainability in ML models. It also presents a detailed explanation of Black-, White-, and Grey-Box models characteristics and the main interpretation techniques and results. Section 3 presents the state of the art methods for interpreting machine learning models. Section 4 presents a description of the proposed Grey-Box classification model. Section 5 presents our datasets utilized in our experiments while Section 6 presents our experimental results. In Section 7, we conducted a detailed discussion of our experimental results and model's method. Finally, Section 8 sketches our conclusions and future research.

2. Interpretability in Machine Learning

2.1. Need of Interpretability and Explainability

As we stated before, the interpretability and the explainability constitute key factors which are of high significance in ML practical models. Interpretability is the ability of understanding and observing a model's mechanism or prediction behavior depending on its input stimulation while, explainability is the ability to demonstrate and explain in understandable terms to a human, the models' decision on predictions problems [7,9]. There are a lot of tasks where the prediction accuracy of a ML model is not the only important issue but, equally important is the understanding, the explanation and the presentation of the model's decision [8]. For example, cancer diagnosis must be explained and justified, in order to support such vital and crucial prediction [10]. Additionally, a medical doctor occasionally must explain his decision for his patient's therapy in order to gain his trust [11]. An employee must be able to understand and explain the results he acquired from the ML models to his boss or co-workers or customers and so on. Furthermore, understanding a ML model can also reveal some of its hidden weaknesses which may lead to further improvement of model's performance [9].

Therefore, an efficient ML model in general must be accurate and interpretable in its predictions. Unfortunately, developing and optimizing an accurate and, at the same time, interpretable model is a very challenging task as there is a "trade-off" between accuracy and interpretation. Most of ML models in order to achieve high performance in terms of accuracy, often become hard to interpret. Kuhn and Johnson [8] stated that "Unfortunately, the predictive models that are most powerful are usually the least interpretable". Optimization of models' accuracy leads to increased complexity, while less complex models with few parameters are easier to understand and explain.

2.2. Black-, White-, Grey-Box Models vs. Interpretability and Explainability

ML models depending on their accuracy and explainability level can be categorized in three main domains: White-Box, Black-Box, and Grey-Box ML models.

A *White-Box* is a model whose inner logic, workings and programming steps are transparent and therefore it's decision making process is interpretable. Simple decision trees are the most common example of White-Box models while other examples are linear regression models, Bayesian Networks and Fuzzy Cognitive Maps [12]. Generally, linear and monotonic models are the easiest models to explain. White-Box models are more suitable for applications which require transparency in their predictions such as medicine and finance [5,12].

In contrast to the White-Box, *Black-Box* is often a more accurate ML model whose inner workings are not known and are hard to interpret meaning that a software tester, for example, could only know the expected inputs and the corresponding outputs of this model. Deep or shallow neural networks are the most common examples of ML Black-Box models. Other examples are support vector machines and also ensemble methods such as boosting and random forests [5]. In general, non-linear and non-monotonic models are the hardest functions to explain. Thus, without being able to fully understand their inner workings it is almost impossible to analyze and interpret their predictions.

A Grey-Box is a combination of Black-Box and White-Box models [13]. The main objective of a Grey-Box model is the development of an ensemble of black and White-Box models, in order to combine and acquire the benefits of both, building a more efficient global composite model. In general, any ensemble of ML learning algorithms containing both black and White-Box models, like neural networks and linear regression, can be considered as a Grey-Box. In a recent research, Grau et al. [12] combined White- and Black-Box ML models following a self-labeled methodology in order to build an accurate and transparent and therefore interpretable prediction model.

2.3. Interpretation Techniques and Results

The techniques for interpreting machine learning models are taxonomized based on the type of the interpretation results we wish or we are able to acquire and on the prediction model framework (White, Black-, or Grey-Box model) we are intending to use.

2.3.1. Two Main Categories

In general, there are two main categories of techniques which provide interpretability in machine learning. These are intrinsic interpretability and post-hoc interpretability [7,14]. *Intrinsic interpretability* is acquired by developing prediction models which are by their nature interpretable, such as all the White-Box models. *Post-hoc interpretability* techniques aim to explain and interpret the predictions of every model, although they are not able to access the model's inner structure and internals, like its weights, in contrast to intrinsic interpretability. For this task, these techniques often develop a second model which has the role of the explanator for the first model. The first model is the main predictor and is probably a Black-Box model without forbidding to be a White-Box model. Local interpretable model-agnostic explanations (LIMEs) [15], permutation feature importance [16], counterfactual explanations [17] and Shapley additive explanations (SHAPs) [18] are some examples of post-hoc interpretability techniques. We conduct a more detailed discussion on these techniques in the following section.

2.3.2. Intrinsic vs. Post-Hoc Interpretability

Intrinsic interpretation methods are by nature interpretable, meaning that there is no need of developing a new model for interpreting the predictions. Thus, the great advantage of these methods is their uncontested and easy achieved interpretation and explanation of models' predictions. On the other hand, their obvious disadvantage is their low performance in terms of accuracy compared to post-hoc methods since the output prediction model on intrinsic methods is a White-Box model.

Post-hoc interpretation methods have the obvious advantage of high performance in terms of accuracy, compared to intrinsic methods, since they often have a Black-Box model as their output predictor, while the interpretation of the predictions is achieved by developing a new model. In contrast, the disadvantage is that developing a new model to explain the predictions of the main Black-Box predictor, is like an expert asking another expert to explain his own decisions, meaning that the explanation of the predictions in these methods cannot come up from the inner workings of the main Black-Box predictor and thus it often needs the access of the original dataset in which the Black-Box model was trained in order to explain its predictions. A further discussion on this subject is conducted in the next section.

2.3.3. Interpretation of Results

The main types of interpretation methods are feature summary statistic, feature summary visualization, model internals, and data points. Feature summary statistics could be a table containing a number per feature, where this number expresses the feature importance. *Feature summary visualization* is another way to present features statistics, instead of tables. By this way the feature summary statistics are visualized in a way to be even more meaningful on the presentation. Model internals are the inner coefficients and parameters of models, such as the weights of all linear models and the learned tree structure of all decision trees. All White-Boxes fall in this category as by nature these models' inner workings are transparent and thus, they are interpreted by "looking" at their internal model parameters. Data points [14] are existing instances or artificially created ones which are used by some methods to interpret a model's prediction. For example, counterfactual explanations falls into this category of methods which explains a model's prediction for a data point by finding similar data points. A further explanation on this subject is conducted in the next section.

3. State of the Art Machine Learning Interpretability Methods

Based on the main interpretation techniques and the types of interpretation results, as mentioned above, there has been a significant research work in the last years for developing sophisticated methods and algorithms in order to interpret machine learning models. These algorithms are always bounded by the trade-off between interpretability and accuracy, meaning that in order to achieve high accuracy there is often sacrifice in interpretability, although the main objective of all these algorithms is the development of both accurate and interpretable models.

LIME [15] can be considered a Grey-Box post-hoc interpretability model which aims to explain the predictions of a Black-Box model by training a White-Box on a local generated dataset, in order to approximate the Black-Box model for the local area on the data points of interest (instances we wish to explain). More specifically, this algorithm generates a fake dataset based on the specific example we wish to explain. This fake dataset contains the perturbed instances of the original dataset and the corresponding predictions of the black model, which are the target values (labels) for this fake dataset. Then, by choosing a neighborhood around the point of interest (define weights of fake data points depending on their proximity to the point of interest) an interpretable model (White-Box) is trained with this new dataset and the prediction is explained by interpreting this local White-Box model. This model does not have to approximate well the Black-Box globally but has to approximate well the Black-Box model locally to the data point of interest. LIME is an algorithm which has the potential to interpret any Black-Box model predictions. However, generating a meaningful fake dataset is a hard and challenging task to do while its great disadvantage lies in the proper definition of the neighborhood around the point of interest [14].

Permutation feature importance [16] is a post-hoc interpretation method which can be applied to Black and White-Box models. This method provides a way to compute features' importance for any Black-Box model by measuring the model's prediction error when a feature has been permuted. A feature will be considered important if the prediction error increases after its permutation and considered as "not important" if the prediction error remains the same. One clear disadvantage of this method is that it requires the actual output of a dataset in order to calculate features' importance. In a case where we only have the trained Black-Box model without having access to the labels of a dataset (unlabeled data), then this method would be unable to interpret the predictions for this Black-Box model.

Counterfactual explanations [17] is a human friendly post-hoc interpretation method which explains a model's prediction by trying to figure out how much the features of an instance of interest must change in order to produce the opposite output prediction. For example, if a Black-Box model predicts that a person is not approved for a bank loan and then by changing his age from "10" to "30" and his job state from "unemployed" to "employed" the model predicts that he is approved for a loan, then this is a counterfactual explanation. In this example, the Black-Box by itself does not give any information or explanation to the person for the loan rejection prediction, while the counterfactual explanation method indicates that if he was older and had a job, he could have taken the loan. One big disadvantage is that this method may find multiple counterfactual explanations for each instance of interest and such an explanation cannot be approved since it would confuse most people.

SHAP [18] is another post-hoc interpretation method which has its roots in game theory and Shapley values. The main objective of this method is the explanation of individual predictions by computing the Shapley values. Shapley value is the contribution of each feature for an instance of interest to the model's prediction. The contribution of each feature is computed by averaging the marginal contributions for all possible permutations for all features. The marginal contribution of a specific feature is the value of the joint contribution for a set of features being this specific feature the last in order, minus the joint contribution for the same set of features with this specific feature being absent. The main innovation of SHAP, is that it computes the Shapley values in a way to minimize the computation time, since all possible permutations for a big number of features would lead to a computationally unattainable task. One disadvantage of this method is that the explanation of a prediction for a new instance requires access to the training data in order to compute the contributions of each feature.

Interpretable mimic learning [19] is a Grey-Box intrinsic interpretation method which trains an accurate and complex Black-Box model and transfers its knowledge to a simpler interpretable model (White-Box) in order to acquire the benefits of both models for developing an accurate and interpretable Grey-Box model. This idea comes from knowledge distillation [20] and mimic learning [21]. More specifically the interpretable mimic learning method trains a Black-Box model. Then the Black-Box model makes predictions on sample dataset and its predictions with this sample dataset are utilized for training a White-Box model. In other words, the target values of the student (White-Box) model, are the output predictions of the teacher (Black-Box) model. One reason that this technique works is that the Black-Box eliminates possible noise and errors in the original training data which could lead to the reduction of the accuracy performance for a White-Box model. This Grey-Box model can have improved performance, comparing to a single White-Box model trained on the original data, being at the same time interpretable since the output predictor is a White-Box model and that is the reason why this model falls in the intrinsic interpretation category. The obvious disadvantage of this method is the same with all intrinsic interpretation models and this is the lower accuracy comparing to post-hoc models, since the performance of these models is bounded by the performance of their output predictor which is a White-Box model.

4. Grey-Box Model Architecture

In this section, we present a detailed description of the proposed Grey-Box model which is based on a self-training methodology.

In our framework, we aim to develop a Grey-Box classification model which will be interpretable, being at the same time nearly as accurate as a Black-Box and more accurate than any single White-Box. We recall that ML models which utilize very complex functions (Black-Box models) are often more accurate but they are difficult and almost impossible to interpret and explain. On the other hand, White-Box ML models solve the problem of explainability but are often much less accurate. Therefore, a Grey-Box model, which is a combination of black and White-Box models, can result in a more efficient global model which possess the benefits of both.

In our model, we utilized an accurate classification model (Black-Box base learner) on a self-training strategy in order to acquire an enlarged labeled dataset which will be used for training an interpretable model (White-Box learner). More specifically, the Black-Box was trained on an initial labeled dataset in order to make predictions on a pool of unlabeled data, while the confidence rate of each prediction was calculated too. Then, the most confident predictions (m.c.p) were added to the initial labeled data and the Black-Box learner was re-trained with the new enlarged labeled data. The same procedure is repeated till a stopping criterion is met. At the end of iterations, the final augmented labeled data are utilized for training the White-Box learner.

An overview of this Grey-Box model architecture and flow chart are depicted in Figures 1 and 2, respectively.

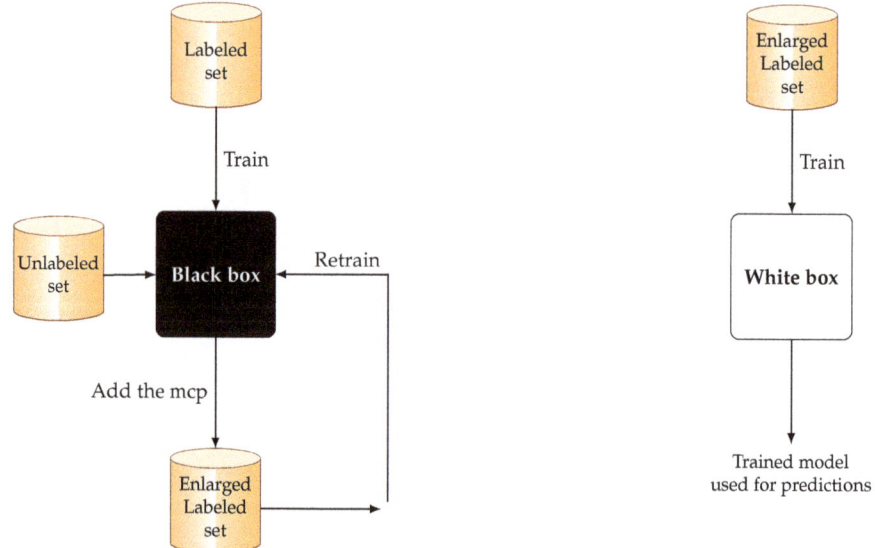

Figure 1. Grey-Box model architecture.

It is worth mentioning that by the term White-Box model we mean that a White-Box learner is used in both left and right architecture displayed in Figure 1, while by Black-Box model we mean that a black box learner is used in both left and right architecture. Additionally, in case the same learner is utilized in both left and right architecture, then the proposed framework is reduced to the classical self-training framework.

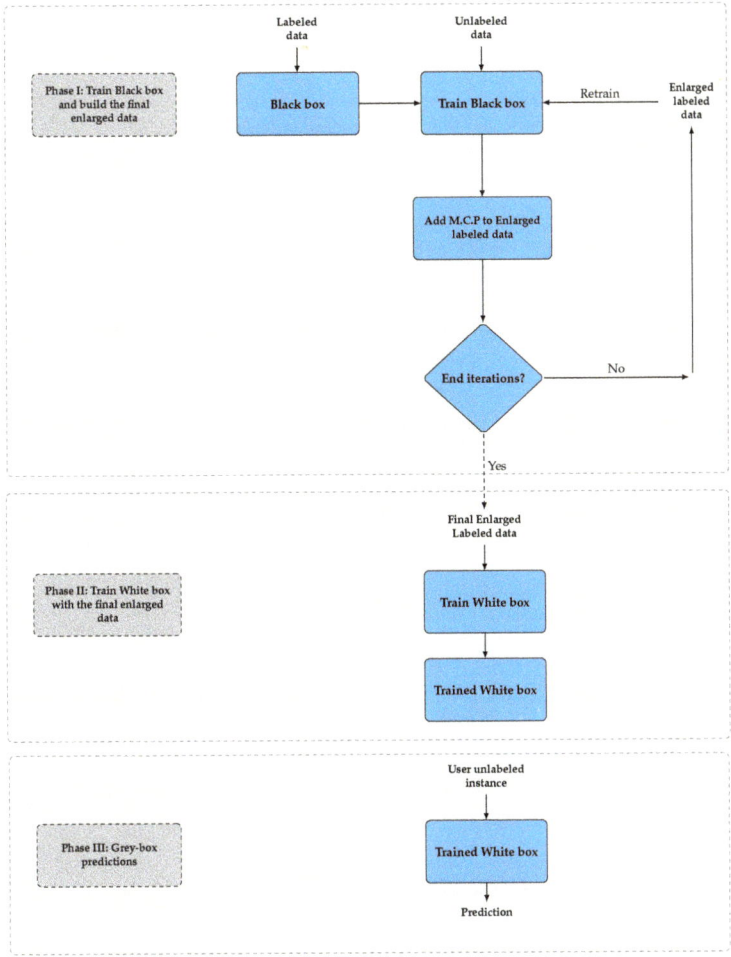

Figure 2. Grey-Box model flow chart.

5. Description of Datasets

In order to evaluate the efficiency and the flexibility of the proposed Grey-Box model, we utilized in our experiments several benchmark datasets on various real-world application domains. These datasets are two educational datasets by a private school, two financial datasets (Australian credit and Bank marketing acquired from UCI Machine Learning Repository) and two medical datasets (Coimbra and Breast Cancer Wisconsin). A brief description of the datasets characteristics and structure is presented in Table 1.

The educational datasets were collected by a Microsoft showcase school during the years 2007–2016 concerning the performance of 2260 students in courses of "algebra" and "geometry" of the first two years of Lyceum [22]. The attributes of this dataset have numerical values ranging from 0 (lowest grade) to 20 (highest grade) referring to the students' performance on the 1st and 2nd semester and the dependent variable of this dataset has four possible states ("fail", "good", "very good", "excellent") indicating the students' grade in the final examinations. Since it is of high importance for an educator to recognize weak students in the middle of the academic period, two datasets have been created, namely EduDataA and EduDataAB. EduDataA contains the attributes concerning the students' performance during the 1st semester. EduDataAB contains the attributes concerning the students' performance during the 1st and 2nd semesters.

The Australian credit dataset [22] concerns approved or rejected credit card applications, containing 690 instances and 14 attributes (six numerical and eight categorical). An interesting thing with this dataset is the variation and the mixture of attributes which is continuous, nominal with small numbers of values and nominal with larger numbers of values. The Bank Marketing dataset [23] is associated with direct marketing campaigns (phone calls) of a Portuguese banking institution. The goal is to identify if the client will subscribe a term deposit (target value y). For the bank marketing staff is very important to know potential clients for subscribing a term deposit since they could decide on which customer to call and which one to leave alone without bothering them with irrelevant product calls. The Bank Marketing dataset consists of 4119 instances and 20 features.

The Coimbra dataset [24] is comprised by ten attributes, all quantitative and a binary dependent variable which indicates the presence or absence of breast cancer. All clinical features were measured for 64 patients with breast cancer and 52 healthy controls. The developing of accurate prediction models based on these ten attributes can potentially be used as a biomarker of breast cancer. The Breast Cancer Wisconsin dataset [22] is comprised by 569 patient instances and 32 attributes. The features were computed from a digitized image of a fine needle aspirate of a breast mass. They describe characteristics of the cell nuclei present in the image. The output attribute is the patient's diagnosis (malignant or benign).

Table 1. Benchmark datasets summary.

Dataset	Application Domain	#Instances	#Features	#Classes
EduDataA	Education	2260	5	4
EduDataAB	Education	2260	10	4
Australian credit	Finance	690	14	2
Bank marketing	Finance	4119	20	2
Coimbra	Medicine	116	10	2
Wisconsin	Medicine	569	32	2

6. Experimental Results

We conducted a series of experimental tests in order to evaluate the performance of the proposed Grey-Box model utilizing the classification accuracy and F_1-score as an evaluation metric with a 10-fold cross-validation.

In these experiments, we compared the performance of Black-box, White-box and Grey-box models using five different labeled dataset ratios i.e., 10%, 20%, 30%, 40%, and 50%, in six datasets. The term "ratio" refers to the ratio of labeled and unlabeled datasets utilized for each experiment. The White-Box model utilizes a single White-Box base learner in self-training framework, the Black-Box model utilizes a single Black-Box base learner in self-training framework while the Grey-Box model utilizes a White-Box learner with a Black-Box base learner in self-training framework.

Additionally, in order to investigate which Black-Box or White-Box combination brings the best results, we tried two White-Box and three Black-Box learners. For this task we have utilized the Bayes-net (BN) [25] and the random tree (RT) [26] as White-Box learners and the sequential minimal optimization (SMO) [27], the random forest (RF) [28], and the multi-layer perceptron (MLP) [29] as Black-Boxes for each dataset. Probably these classifiers are the most popular and efficient machine learning methods for classification tasks [30]. Table 2 presents a summary of all models used in the experiments.

The experimental analysis was conducted in a two-phase procedure: In the first phase, we compared the performance of Grey-Box models against that of White-Box models, while in the second phase, we evaluated the prediction accuracy of the Grey-Box models against that of Black-Box models.

Table 2. White-, Black-, and Grey-Box models summary.

Model	Description
White-Box (BN)	base learner: BN
White-Box (RT)	base learner: RT
Black-Box (SMO)	base learner: SMO
Black-Box (RF)	base learner: RF
Black-Box (MLP)	base learner: MLP
Grey-Box (SMO- BN)	base learners: SMO, BN
Grey-Box (RF- BN)	base learners: RF, BN
Grey-Box (MLP- BN)	base learners MLP, BN
Grey-Box (SMO-RT)	base learners: SMO, RT
Grey-Box (RF-RT)	base learners: RF, RT
Grey-Box (MLP-RT)	base learners: MLP, RT

6.1. Grey-Box Models vs. White-Box Models

Figures 3–8 present the performance of the Grey-Box and White-Box models. Regarding both metrics we can observe that the Grey-Box model outperformed the White-Box model for each Black-Box and White-Box classifier combination in most cases, especially on EduDataA, EduDataAB, and Australian datasets. On Coimbra dataset the two models exhibit almost similar performance, with Grey-Box model being slightly better. On Winsconsin dataset the Grey-Box model reported better performance in general, except for ratio 40 and 50 where the White-Box model outperformed the Grey-Box, utilizing RT as White-Box base learner. The interpretation of Figures 3–8 reveals that for EduDataA, EduDataAB and Bank marketing datasets, as ratio value increases, the performance of all models increases too, while on the rest datasets, it seems that we cannot state a similar conclusion. We can also observe that the Grey-Box model performs better in most cases for small ratios (10%, 20%, 30%) while it seems that for higher ratios (40%, 50%) the White-Box exhibits similar performance as the Grey-Box model. Summarizing, we point out that the proposed Grey-Box model performs better for small labeled ratios. Additionally, the Grey-Box (SMO-RT) reported the best overall performance reporting the highest classification accuracy in three out of six datasets, while the Grey-Box reported the best performance utilizing RT as White-Box learner, regarding all utilized ratios.

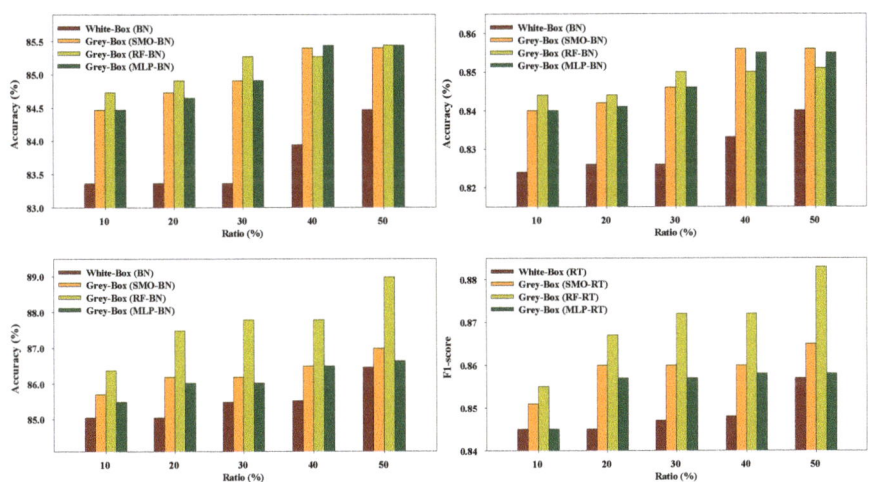

Figure 3. Box-plot with performance evaluation of Grey-Box and White-Box models for EduDataA dataset.

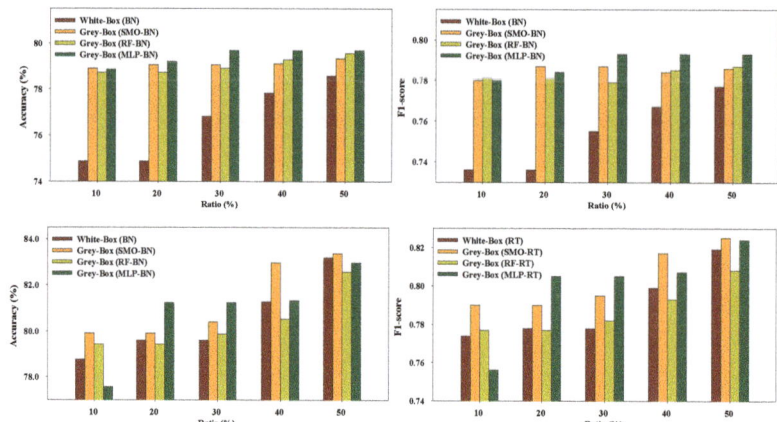

Figure 4. Box-plot with performance evaluation of Grey-Box and White-Box models for EduDataAB dataset.

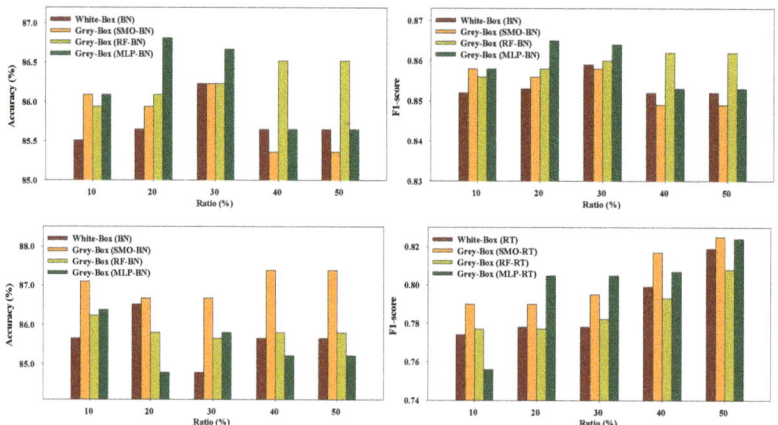

Figure 5. Box-plot with performance evaluation of Grey-Box and White-Box models for Australian dataset.

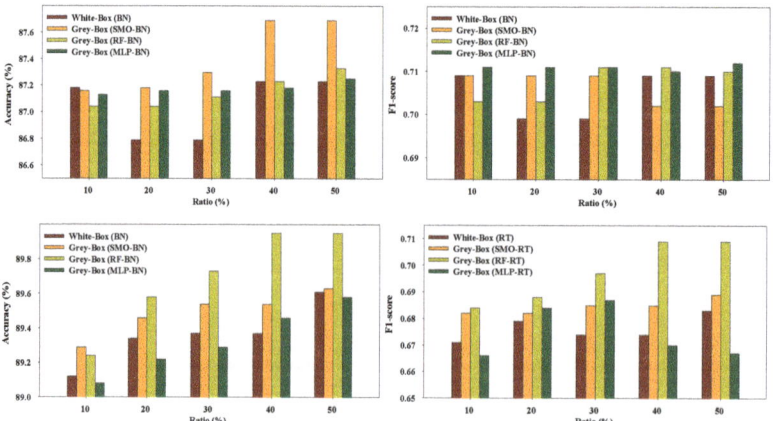

Figure 6. Box-plot with performance evaluation of Grey-Box and White-Box models for Bank dataset.

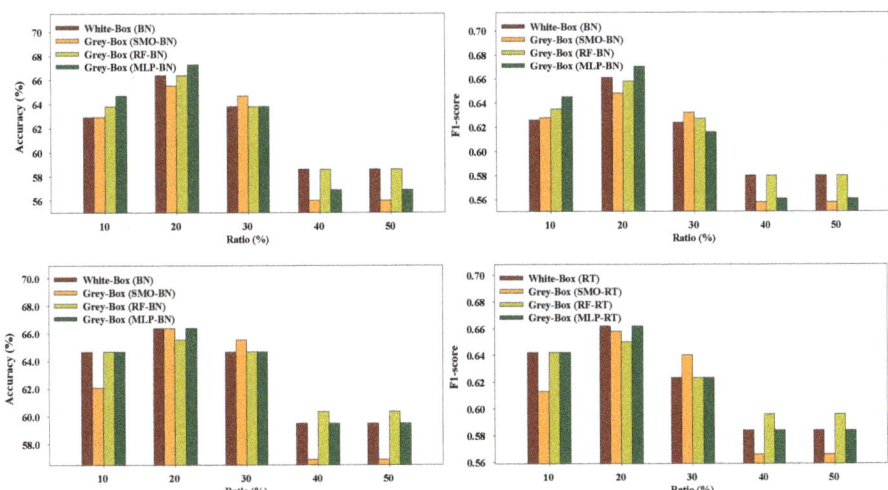

Figure 7. Box-plot with performance evaluation of Grey-Box and White-Box models for Coimbra dataset.

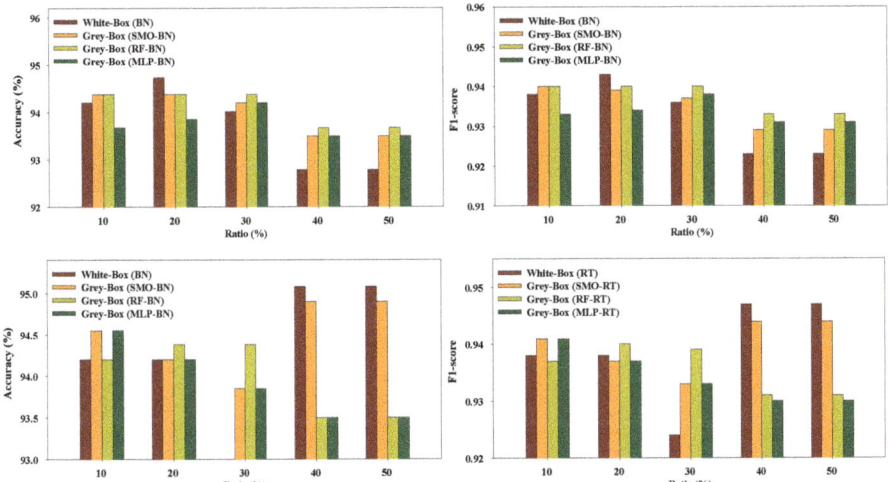

Figure 8. Box-plot with performance evaluation of Grey-Box and White-Box models for Winsconsin dataset.

6.2. Grey-Box Models vs. Black-Box Models

Figures 9–14 present the performance of the Grey-Box and Black-Box models utilized in our experiments. We can conclude that the Grey-Box model was nearly as accurate as a Black-Box model with the Black-Box being slightly better. More specifically, the Black-Box model clearly outperformed the Grey-Box model on EduDataA and EduDataAB datasets while the Grey-Box reported better performance on Australian dataset. On Bank marketing, Coimbra, and Winsconsin datasets both models exhibited similar performance.

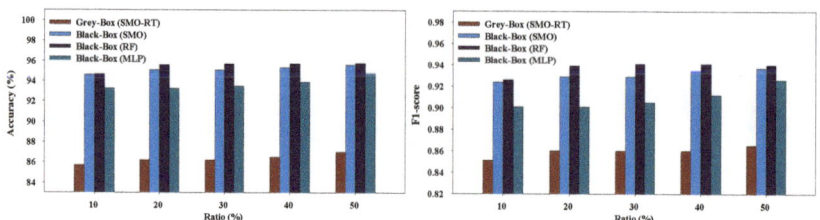

Figure 9. Box-plot with performance evaluation of Grey-Box and Black-Box models for EduDataA dataset.

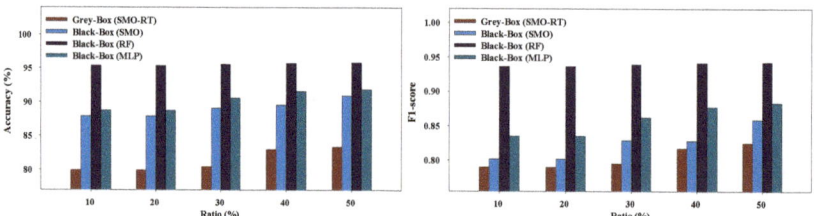

Figure 10. Box-plot with performance evaluation of Grey-Box and Black-Box models for EduDataAB dataset.

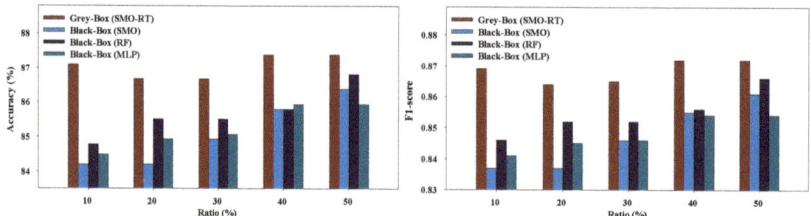

Figure 11. Box-plot with performance evaluation of Grey-Box and Black-Box models for Australian dataset.

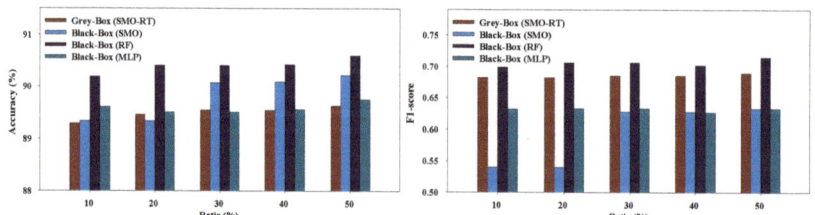

Figure 12. Box-plot with performance evaluation of Grey-Box and Black-Box models for Bank dataset.

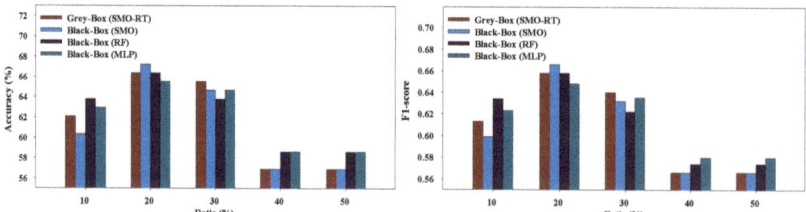

Figure 13. Box-plot with performance evaluation of Grey-Box and Black-Box models for Coimbra dataset.

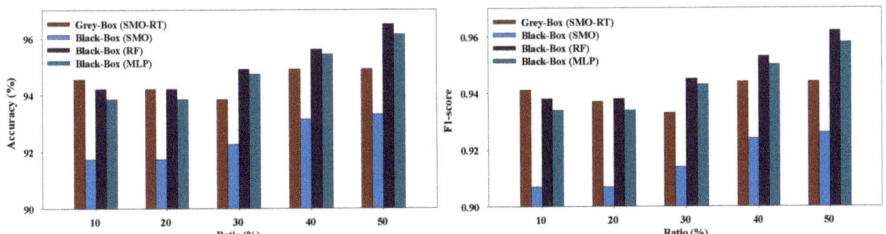

Figure 14. Box-plot with performance evaluation of Grey-Box and Black-Box models for Winsconsin dataset.

7. Discussion

7.1. Discussion about the Method

In this work, we focus on developing an interpretable and accurate Grey-Box prediction model applied on datasets with a small number of labeled data. For this task, we utilized a Black-Box base learner on a self-training methodology in order to acquire an enlarged labeled dataset which was used for training a White-Box learner. Self-training was considered the most appropriate self-labeled method to use, due to its simplicity, efficiency and because is a widely used single view algorithm [4]. More specifically, in our proposed framework, an arbitrary classifier is initially trained with a small amount of labeled data, constituting its training set which is iteratively augmented using its own most confident predictions of the unlabeled data. More analytically, each unlabeled instance which has achieved a probability over a specific threshold is considered sufficiently reliable to be added to the labeled training set and subsequently the classifier is retrained. In contrast to the classic self-training framework, we utilized the White-Box learner as the final predictor. Therefore, by training a Black-Box model on the initial labeled data we obtain a larger and a more accurate final labeled dataset, while utilizing a White-Box trained on the final labeled data and being the final predictor, we are able to obtain explainability and interpretability on the final predictions.

In contrast to Grau et al. [12] approach, we decided to augment the labeled dataset utilizing the most confident predictions from the unlabeled dataset. This way we may lose some extra instances, leading to a smaller final labeled dataset, but instead we manage to avoid mislabeling of new instances and thus reduce the "noise" of the new augmented dataset. Thus, we addressed the weakness of including erroneous labeled instances in the enlarged labeled dataset which could probably lead the White-Box classifier to poor performance.

Che et al. [19] proposed an interpretable mimic model which falls into the intrinsic interpretability category similar to our proposed model. Nevertheless, the main difference compared to our approach is that we aim to add interpretability on semi supervised algorithms (self-training) which means that our proposed model is mostly recommended on scenarios in which the labeled dataset is limited while a big pool of unlabeled data is available.

7.2. Discussion about the Prediction Accuracy of the Grey-Box Model

In terms of accuracy, our proposed Grey-Box model outperformed White-Box model in most cases of our experiments especially on our educational datasets. As regards, Coimbra dataset, the two models reported almost identical performance with Grey-Box to be slightly better. Relative to Winsconsin dataset there were two cases in which the White-Box model reported better performance over the Grey-Box. Probably this occurs due to the fact that these datasets are very small (few number of labeled instances) and becoming even smaller by the self-training framework as the datasets have to be split in labeled and unlabeled datasets.

Most of Black-Box models are by nature "data-eaters", especially neural networks, meaning that they need a large number of instances in order to outperform most of White-Box models in terms of accuracy. Our Grey-Box model consists of a Black-Box base learner which was initially trained on the labeled dataset and then makes predictions on the unlabeled data in order to augment the labeled dataset which is then used for training a White-Box learner. If the initial dataset is too small, there is the risk that the Black-Box could exhibit poor performance and make wrong predictions on the unlabeled dataset. This implies that the final augmented dataset will have a lot of mislabeled instances forcing the White-Box, which is the output of the Grey-Box model, to exhibit low performance too. As a result, the proposed Grey-Box model may have lower or equal performance compared to the White-Box model.

Moreover, the Grey–Box model performed slightly worse than the Black-Box models as expected since our goal was to demonstrate that the proposed Grey-Box model achieved a performance almost as good as a Black-Box model. Thus, in summary our results reveal that our model exhibited better performance compared to the White-Box model and reported comparable performance to the Black-Box model.

7.3. Discussion about the Interpretability of the Grey-Box Model

The proposed Grey-Box model falls into the intrinsic interpretability category as its output predictor is a White-Box model and by nature all White-Box models are interpretable. Nevertheless, it does not have the disadvantage of low accuracy inherent to methods of this category as it performs as good as post-hoc methods.

Also, in order to present how the interpretability and explainability of our Grey-Box model comes up in practice, we conducted a case study (experiment example) for the Coimbra dataset utilizing Grey-Box (SMO-RT) because this model exhibited better overall performance. For this task, we made predictions for two instances (persons) utilizing our trained Grey-Box model and now we attempt to explain, in understandable to every human terms, why our model made these predictions. Table 3 presents the values of each attribute for two instances of the Coimbra dataset while a detailed description of the attributes can be found in [24].

Table 3. Feature values for two persons from the Coimbra dataset.

	Features									
	Age	BMI	Glucose	Insulin	HOMA	Leptin	Adiponectin	Resistin	MCP-1	Class
Instance1	51	22.892	103	2.74	0.696	8.016	9.349	11.554	359.232	Patient
Instance2	36	28.570	86	4.34	0.920	15.12	8.600	9.150	534.22	Healthy

Figure 15 presents the trained Grey-Box model's inner structure. From the interpretation of the Grey-Box model, we conclude that the model suggests as the most significant feature, the "age", since it is in the root of the tree, while we can also observe that a wide range of features are not utilized at all from the model's inner structure.

Furthermore, by observing the inner structure of the model we can easily extract the model predictions for both instances by just following the nodes and the arrows of the tree, based on the feature values of each person, this procedure is performed, until we reach a leaf of the tree which has the output prediction of the model. More specifically, the Instance1 has "Age = 51" and for the first node the statement "age $\leq$ 37" is "false", thus we move right to the next node which has the statement "HOMA $\leq$ 2.36". This is "true", so we move left to "glucose $\leq$ 91.5" which is "false" since the Instance1 has "glucose = 103". Therefore, the model's prediction for the Instance1 is "patient". By a similar procedure we can extract the model's prediction for Instance2 which is "healthy". Next, we can also draw some rules from the model and translate them into explanations such as

- Rule 1: "The persons which are of age over 37, have HOMA lower than 2.36, and have glucose over than 91.5, most probably are patients".
- Rule 2: "The persons which are of age lower than 37 and have glucose lower than 91.5 most probably are healthy."

and so on. Instance1 falls exactly on Rule 1, while Instance2 falls exactly on Rule 2. Thus, an explanation for the Grey–Box model's predictions could be that Instance1 is probably "patient" because he/she is older than 37 years old and has very low HOMA with high glucose level, while Instance2 is probably "healthy" because he/she is under 37 years old and has low glucose level.

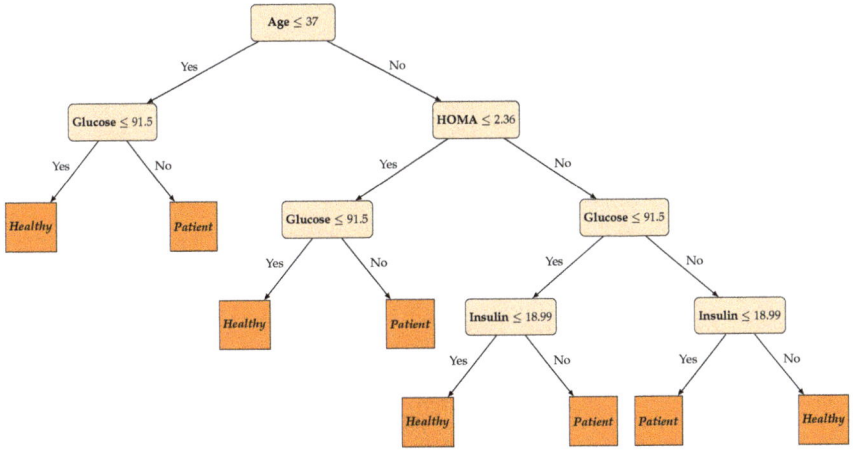

Figure 15. The interpretation of the learned inner model structure.

8. Conclusions

In this work, we developed a high performance and interpretable Grey-Box ML model. Our main purpose was to acquire both benefits of black and White-Box models in order to build an interpretable classifier with a better classification performance, compared to a single White-Box model and comparable to that of a Black-Box. More specifically, we utilized a Black-Box model to enlarge an initial labeled dataset and the final augmented dataset was used to train a White-Box model. In contrast to the self-training framework, this trained White-Box model was utilized as the final predictor. This ensemble model falls into the category of intrinsic interpretability since its output predictor is a White-Box which is by nature interpretable. Our experimental results revealed that our proposed Grey-Box model has accuracy comparable to that of a Black-Box but better accuracy comparing to a single White-Box model, being at the same time interpretable as a White-Box model.

In our future work, we aim to extend our experiments of the proposed model to several datasets and improve its prediction accuracy with more sophisticated and theoretically motivated ensemble learning methodologies combining various self-labeled algorithms.

Author Contributions: E.P., I.E.L. and P.P. conceived of the idea, designed and performed the experiments, analyzed the results, drafted the initial manuscript and revised the final manuscript. All authors have read and agreed to the published version of the manuscript.

Funding: This research received no external funding.

Conflicts of Interest: The authors declare no conflict of interest.

References

1. Chapelle, O.; Scholkopf, B.; Zien, A. Semi-supervised learning. *IEEE Trans. Neural Netw.* **2009**, *20*, 542. [CrossRef]
2. Zhu, X.; Goldberg, A.B. Introduction to semi-supervised learning. *Synth. Lect. Artif. Intell. Mach. Learn.* **2009**, *3*, 1–130. [CrossRef]
3. Livieris, I.E. A new ensemble self-labeled semi-supervised algorithm. *Informatica* **2019**, *43*. [CrossRef]

4. Triguero, I.; García, S.; Herrera, F. Self-labeled techniques for semi-supervised learning: Taxonomy, software and empirical study. *Knowl. Inf. Syst.* **2015**, *42*, 245–284. [CrossRef]
5. Robnik-Šikonja, M.; Kononenko, I. Explaining classifications for individual instances. *IEEE Trans. Knowl. Data Eng.* **2008**, *20*, 589–600.
6. Bratko, I. Machine learning: Between accuracy and interpretability. In *Learning, Networks and Statistics*; Springer: Berlin, Germany, 1997; pp. 163–177.
7. Arrieta, A.; Díaz-Rodríguez, N.; Del Ser, J.; Bennetot, A.; Tabik, S.; Barbado, A.; García, S.; Gil-López, S.; Molina, D.; Benjamins, R.; et al. Explainable Artificial Intelligence (XAI): Concepts, Taxonomies, Opportunities and Challenges toward Responsible AI. *Inf. Fusion* **2020**, *58*, 82–115. [CrossRef]
8. Kuhn, M.; Johnson, K. *Applied Predictive Modeling*; Springer: New York, NY, USA, 2013; Volume 26.
9. Doshi-Velez, F.; Kim, B. Towards a rigorous science of interpretable machine learning. *arXiv* **2017**, arXiv:1702.08608.
10. Holzinger, A.; Kieseberg, P.; Weippl, E.; Tjoa, A.M. Current advances, trends and challenges of machine learning and knowledge extraction: From machine learning to explainable ai. In Proceedings of the International Cross-Domain Conference for Machine Learning and Knowledge Extraction, Hamburg, Germany, 27–30 August 2018; Springer: Berlin, Germany, 2018; pp. 1–8.
11. Samek, W.; Wiegand, T.; Müller, K.R. Explainable artificial intelligence: Understanding, visualizing and interpreting deep learning models. *arXiv* **2017**, arXiv:1708.08296.
12. Grau, I.; Sengupta, D.; Garcia, M.; Nowe, A. Grey-Box Model: An ensemble approach for addressing semi-supervised classification problems. In Proceedings of the 25th Belgian-Dutch Conference on Machine Learning BENELEARN, Kortrijk, Belgium, 12–13 September 2016.
13. Bohlin, T.P. *Practical Grey-Box Process Identification: Theory and Applications*; Springer Science & Business Media: Berlin/Heidelberg, Germany, 2006.
14. Molnar, C. *Interpretable Machine Learning: A Guide for Making Black Box Models Explainable (eBook)*; Lulu: Morrisville, NC, USA, 2019.
15. Ribeiro, M.T.; Singh, S.; Guestrin, C. Why should I trust you? Explaining the predictions of any classifier. In Proceedings of the 22nd ACM SIGKDD International Conference on Knowledge Discovery and Data Mining, San Francisco, CA, USA, 13–17 August 2016; ACM: New York, NY, USA, 2016; pp. 1135–1144.
16. Fisher, A.; Rudin, C.; Dominici, F. Model class reliance: Variable importance measures for any machine learning model class, from the "Rashomon" perspective. *arXiv* **2018**, arXiv:1801.01489.
17. Wachter, S.; Mittelstadt, B.; Russell, C. Counterfactual Explanations without Opening the Black Box: Automated Decisions and the GPDR. *Harv. JL Tech.* **2017**, *31*, 841. [CrossRef]
18. Lundberg, S.M.; Lee, S.I. A unified approach to interpreting model predictions. In Proceedings of the Advances in Neural Information Processing Systems, Long Beach, CA, USA, 4–9 December 2017; pp. 4765–4774.
19. Che, Z.; Purushotham, S.; Khemani, R.; Liu, Y. Interpretable deep models for ICU outcome prediction. In *AMIA Annual Symposium Proceedings*; American Medical Informatics Association: Bethesda, MD, USA, 2016; Volume 2016, p. 371.
20. Hinton, G.; Vinyals, O.; Dean, J. Distilling the knowledge in a neural network. *arXiv* **2015**, arXiv:1503.02531.
21. Ba, J.; Caruana, R. Do deep nets really need to be deep? In *Proceedings of the Advances in Neural Information Processing Systems*; Curran Associates, Inc.: New York, NY, USA, 2014; pp. 2654–2662.
22. Bache, K.; Lichman, M. *UCI Machine Learning Repository*; Department of Information and Computer Science, University of California: Irvine, CA, USA, 2013.
23. Moro, S.; Cortez, P.; Rita, P. A data-driven approach to predict the success of bank telemarketing. *Decis. Support Syst.* **2014**, *62*, 22–31. [CrossRef]
24. Patrício, M.; Pereira, J.; Crisóstomo, J.; Matafome, P.; Gomes, M.; Seiça, R.; Caramelo, F. Using Resistin, glucose, age and BMI to predict the presence of breast cancer. *BMC Cancer* **2018**, *18*, 29.
25. Koski, T.; Noble, J. *Bayesian Networks: An Introduction*; John Wiley & Sons: Hoboken, NJ, USA, 2011; Volume 924.
26. Criminisi, A.; Shotton, J. *Decision Forests for Computer Vision and Medical Image Analysis*; Springer Science & Business Media: Berlin/Heidelberg, Germany, 2013.
27. Bentaouza, C.; Benyettou, M.; Abe, S.; Alphonso, I.; Bishop, C.; Boser, B.; Guyon, I.; Vapnik, V.; Brucher, J.; Chatel, M.; et al. Support Vector Machines: Theory and Application. *J. Appl. Sci.* **2005**, *10*, 144.

28. Hartshorn, S. *Machine Learning with Random Forests and Decision Trees: A Visual Guide for Beginners*; Amazon: Seattle, WA, USA, 2016.
29. Demuth, H.B.; Beale, M.H.; De Jess, O.; Hagan, M.T. *Neural Network Design*; Oklahoma State University: Oklahoma, OK, USA, 2014.
30. Wu, X.; Kumar, V.; Quinlan, J.; Ghosh, J.; Yang, Q.; Motoda, H.; McLachlan, G.; Ng, A.; Liu, B.; Yu, P.; et al. Top 10 algorithms in data mining. *Knowl. Inf. Syst.* **2008**, *14*, 1–37. [CrossRef]

© 2020 by the authors. Licensee MDPI, Basel, Switzerland. This article is an open access article distributed under the terms and conditions of the Creative Commons Attribution (CC BY) license (http://creativecommons.org/licenses/by/4.0/).

Article

A Soft-Voting Ensemble Based Co-Training Scheme Using Static Selection for Binary Classification Problems

Stamatis Karlos *, Georgios Kostopoulos and Sotiris Kotsiantis

Educational Software Development Laboratory (ESDLab), Department of Mathematics, University of Patras, 26504 Patras, Greece; kostg@sch.gr (G.K.); sotos@math.upatras.gr (S.K.)
* Correspondence: stkarlos@upatras.gr

Received: 1 November 2019; Accepted: 13 January 2020; Published: 16 January 2020

Abstract: In recent years, a forward-looking subfield of machine learning has emerged with important applications in a variety of scientific fields. Semi-supervised learning is increasingly being recognized as a burgeoning area embracing a plethora of efficient methods and algorithms seeking to exploit a small pool of labeled examples together with a large pool of unlabeled ones in the most efficient way. Co-training is a representative semi-supervised classification algorithm originally based on the assumption that each example can be described by two distinct feature sets, usually referred to as views. Since such an assumption can hardly be met in real world problems, several variants of the co-training algorithm have been proposed dealing with the absence or existence of a naturally two-view feature split. In this context, a Static Selection Ensemble-based co-training scheme operating under a random feature split strategy is outlined regarding binary classification problems, where the type of the base ensemble learner is a soft-Voting one composed of two participants. Ensemble methods are commonly used to boost the predictive performance of learning models by using a set of different classifiers, while the Static Ensemble Selection approach seeks to find the most suitable structure of ensemble classifier based on a specific criterion through a pool of candidate classifiers. The efficacy of the proposed scheme is verified through several experiments on a plethora of benchmark datasets as statistically confirmed by the Friedman Aligned Ranks non-parametric test over the behavior of classification accuracy, F_1-score, and Area Under Curve metrics.

Keywords: binary classification; co-training; ensemble methods; feature views; dynamic ensemble selection; Soft-Voting

---

1. Introduction

In recent years, the latest research on machine learning (ML) which has placed much emphasis on learning from both labeled and unlabeled examples is mainly expressed by semi-supervised learning (SSL) [1]. SSL is increasingly being recognized as a burgeoning area embracing a plethora of efficient methods and algorithms seeking to exploit a small pool of labeled examples together with a large pool of unlabeled ones in the most efficient way. Since in most real-world applications there is an abundance of unlabeled examples, while labeled examples are either difficult or expensive to obtain, SSL has emerged as a promising domain with important applications in a variety of scientific fields with substantial results [2,3].

In general, SSL methods are commonly divided into two key tasks, as follows: semi-supervised classification (SSC) for discrete-value output variables and semi-supervised regression (SSR) for real-value ones [4]. The classification task, usually referred to as pattern recognition in engineering or discriminant analysis in statistics [5], has been widely studied under the semi-supervised framework for classifying any given example to one out of the included class labels into a predetermined set.

Depending on the number of class labels, classification problems may be either binary (with two class labels) or multi-class (with more than two class labels). For that purpose, a number of semi-supervised algorithms have been developed and successfully implemented, such as self-training [6], co-training [7], and tri-training [8], as well as approaches that are based on semi-supervised support vector machines and transductive learning or SSL graph-based methods, to name just a few [9].

Co-training is a representative multi-view SSC algorithm originally based on the assumption that each example can be described by two distinct feature sets, usually referred to as views [7]. Then, two classification algorithms are trained separately on each view and the most confident predictions of each one on the unlabeled data are used to augment the training set of the other. Let L_D denote a small set of labeled examples and U_D a large set of unlabeled ones. The two separate classifiers C_1, C_2 are retrained on the enlarged set L_D and the process is repeated for a predefined number of iterations or until a stopping criterion is satisfied, such as the U_D pool to be empty. However, since such a two-view assumption can hardly be met in real world problems, several variants of the co-training algorithm have been proposed dealing with the absence or existence of a naturally two-view feature split with notable results.

In addition to these, ensemble learning or committee-based learning or learning multiple classifier systems has emerged recently and is considered as one of the most adequate solutions for building powerful and accurate classification models [10]. Instead of using one algorithm for building a learning model, ensemble methods are commonly used to construct and combine a set of classifiers, either weak or strong, generally called base learners. The fundamental points for the effectiveness of an ensemble method concern careful selection of both base learners [11] and the combination method for producing the final hypothesis [12]. Averaging (simple or weighted [13]) and voting (majority, unanimity, plurality, or even weighted votes) are popular and commonly used combination methods [14] depending on the problem which needs to be resolved [10]. Moreover, approaches using committees of base learners into the core of their learning process have also been demonstrated recently, presenting encouraging results [15].

The necessity of accurate and robust decisions inside a semi-supervised scheme play a cardinal role, especially in cases where the number of initially labeled instances is quite small and no decision correcting or editing mechanisms have been placed inside the learning kernel. Therefore, strategies trying to build an ensemble optimizing well-defined criteria could be a useful asset of a compact semi-supervised algorithm for selecting the most suitable structure per task. Although this concept has been highly exploited under the supervised mode, only few works have been detected in the related literature that apply similar approaches [16,17]. Furthermore, there are some works in this field that apply mechanisms suitably combining the decisions of selected base learners, failing, however, to state the reasons of their choice, apart from some generic properties, such as the combination of one generative and one discriminative approach that favors the diversity of the applied co-training algorithm [18].

In this context, a soft-Voting ensemble-based co-training scheme using static selection strategy, regarding binary classification problems, is proposed. Since co-training is primarily relying on the multi-view assumption, a heuristic scheme is adopted for manually generating the two views. Although this random split may act towards injecting diversity into a multi-view SSL approach, the main asset of the proposed algorithm is the construction of an ensemble learner choosing among five different classifiers, based on a novel objective function that measures the efficacy of any examined pair of classifiers per dataset, operating under a soft-Voting scheme. The efficacy of the proposed mechanism for selecting the ensemble's participants is verified through several experiments against both single-view SSL variants—through the well-known self-training scheme—and the co-training scheme, applying all of the 10 different pairs of algorithms into the same soft-Voting learner and the five individual classifiers, on a plethora of benchmark datasets over five separate labeled ratio values. The obtained results regarding two well-known classification metrics are statistically confirmed by the applied Friedman Aligned Ranks non-parametric test.

The rest of this paper is organized as follows: the co-training framework is presented in Section 2, while reviewing recent studies concerning both the application of co-training in real world applications and Ensemble Selection strategies. In Section 3, we propose and describe in detail the proposed co-training scheme operating under a random feature split using internally a static selection strategy regarding a soft-Voting algorithm. Section 4 includes the experiments carried out along with the relevant results. Finally, in Section 5 we comment on the results considering some thoughts for future work.

2. Related Works

This section consists of two different parts, which highlight the two main points related to our proposed work. After having mentioned some of the most important works towards these directions, we can summarize our main contributions into the next section, facilitating the structure of this work.

2.1. Co-Training Studies

The first part of this section is dedicated to Co-training scheme and some of the numerous variants that have been demonstrated. Hence, Co-training is deemed to be a representative multi-view SSL method established by Blum and Mitchell [7] for binary classification problems and, in particular, for categorizing web pages either as course or as non-course. It is based on the premise that each example can be naturally divided into two separate set of features usually referred to as views, which is clearly an assumption of great importance for the implementation of the particular method. Co-training is identified as a "two-view weakly supervised algorithm" [6] since it incorporates the self-training approach to separately teach each one of the two supervised classifiers in the corresponding feature view and boost the classification performance exploiting the unlabeled examples in the most efficient manner [19]. Moreover, Zhu and Goldberg consider co-training as a wrapper method which is not affected by the two supervised classifiers employed in the relevant procedure, provided that they produce good predictions on unlabeled data [20]. Several modifications have been implemented since then, including mutual-learning and co-EM—bringing together the co-training and Expectation-Maximization (EM) approaches—exploiting mainly simple classifiers like naive Bayes (NB) [7,21].

In addition to the "two view" assumption, the effectiveness of the particular method depends largely on two other key assumptions: the first one is that each view is adequate for classifying the unlabeled data using a small set of labeled examples for training, while the second one is that each view is conditionally independent given the class label. When either of these assumptions is not met, different co-training variants have been proposed with comparable results. In the case where the "two view" assumption is not fulfilled, a random feature partition could take place to facilitate the application of the method as proposed by Zhu and Goldberg [20]. In such cases, the feature set is partitioned into two subsets of almost equal size, which henceforth form the two feature views, while different classifiers C_1, C_2 are employed. In addition, the same classifiers may be used under different configuration parameters, thus ensuring the diversity between them [22].

The number of studies that propose co-training as an effective SSL method is really restricted. One of these is presented in [22], where sentiment analysis is the main focus, while in [23] the authors have tackled a health care issue. Although the popularity of this type of problem is widespread and even though any shortcomings that may be associated with a large amount of labeled data can be efficiently leveraged by other SSL methods, yet co-training seems to not have been delved into thoroughly enough. In this study, three different sources of text data were examined: news articles, online reviews, and blogs. A number of co-training variants were designed, focusing on the way the split of the feature space takes place, fitting appropriately the specific properties that characterize text data, such as the creation of one view by unigrams and the rest by bigrams or by adopting character-based language models and bag-of-words models, respectively. The produced results demonstrate the effectiveness of the co-training algorithm.

Another task that has been efficiently tackled by using the co-training method is that of drug discovery, where classification methods need to be applied so as to predict the suitability of some molecules considering treatments of diseases and their possibly induced side-effects during the initial steps of tedious experiments [24]. Accurate predictions may save both time and money, since fewer combinations would be investigated and the final results could be acquired much faster. In this work, two different views were available, stemming from chemistry and biology, and had to be mixed to reach the final conclusion. The approaches that were examined may be summed up as follows: (i) access separately each view either with a base classifier or the partial least squares (PLS) regression method [25], (ii) fuse the different views, either by joining the heterogeneous data without any preprocess or after having applied the PLS method, also used for dimensionality reduction, and (iii) a modification of the co-training method (co-FTF). Ensemble tree-based learners were preferred in this last approach, handling imbalanced datasets appropriately and leading to promising results, while examining two labeled ratio scenarios. In addition, a random forest of predictive clustering trees was incorporated in a self-training scheme for multi-target regression, thus improving the performance of the employed SSL approach [26].

An expansion of the co-training algorithm, which includes an ensemble of tree-based learners as base learner, has been proposed in [27]. Under the assumptions that are presented there, the necessity of two sufficient and redundant views has been eliminated for the proper operation of Co-Forest. Furthermore, the bootstrap method that is exploited during the creation of the included decision trees provides the required diversity and, at the same time, reduces the chance of exporting biased decisions, leading to an efficient operation of the SSL scheme. Adaptive Data Editing based Co-Forest (ADE-Co-Forest) [28] constitutes a variant of the original Co-Forest algorithm, introducing an internal mechanism in order to tackle the mislabeled instances, thus improving the total predictive behavior, since both false negative/positive error rates are further reduced, compared to its ancestor. A boosted co-training algorithm has also been proposed for a real-task problem—to be more specific, it concerns the human action recognition—which is based on the mutual information and the consistency between labeled and unlabeled data. Two metrics, named inter-view and intra-view confidence, are introduced and exploited dynamically so as to select the most appropriate subset of the unlabeled pool with the corresponding pseudo-labels [29].

Recently, a quite effective co-training method was introduced in [30] for early prognosis of undergraduate students' performance in the final examinations of a distance learning course based on attributes which are naturally divided into two separate and independent views. The first one concerns students' characteristics and academic achievements which are manually filled out by tutors, while the second one refers to attributes tracking students' online activity in the course learning management system and which are automatically recorded by the system. It should be mentioned that semi-supervised multi-view learning has also been successfully applied for gene network reconstruction combining the interactions predicted by a number of different inference methods [19]. In a similar work, an ensemble-based SSL approach has been proposed for the computational discovery of miRNA regulatory networks from large-scale predictions produced by different algorithms [31].

2.2. Ensemble Selection Strategies

The second part is oriented towards reporting briefly some of the most important points related with Ensemble Selection concept [32,33]. To be more specific, some usual keywords in this field are Multiple Classification Systems (MCSs), Static Ensemble Classifier (SEC), and Dynamic Ensemble Classifier (DEC), as well as classifiers' competence and diversification. The way that all these terms are connected is the fact that when a new ensemble learner is designed, the main ambitions are the employment of complementary and diverse participants, following the main asset of MCSs regarding the continuous increase of the predictive rate. The main difference between the remaining two terms is the fact that SEC strategies examine a global solution regarding the total set of unknown instances, while the DES approaches provide a separate solution per test instance using mainly local restrictions. Despite their distinct roles, they can be combined under hybrid mechanisms sharing similar measurement metrics or ML techniques for converging to their decisions [34,35].

Ensemble Selection has been inserted as a new stage into the original chain of constructing an ensemble learner, taking into consideration both the importance of computational needs that arise when we trust ensembles with too many participants and the fact of discarding less accurate models or models that reduce the internal diversity. This tactic is usually referred to as ensemble pruning or selective ensemble. A taxonomy of these techniques has been proposed in [36], assigning them to four different categories: (i) ranking-based, (ii) clustering-based, (iii) optimization-based, and (iv) others, including the remaining techniques that cannot be strictly categorized to any of the previous three subsets. Another taxonomy was demonstrated in 2014, concerning mainly the actual need of DES in practice and the relation between the inherent complexity of classification problem, measured by appropriate metrics, and the contribution of the examined Dynamic Selection approaches [16]. Prototype selection techniques have also been examined in the abovementioned framework, acting beneficially towards both reducing computational resources and boosting the classification accuracy [37]. Furthermore, one related work on the field of SSL has been proposed using the competence of selected classifiers that stems from an affinity graph, achieving smoothness of the decisions for neighboring data [17].

3. The Proposed Co-Training Scheme

Motivated by the above studies, in the present paper we make an attempt to put forward an ensemble-based co-training scheme for binary classification problems adopting a strategy of choosing the base classifiers of the ensemble from an available pool of candidate classification algorithms per dataset. The most important points concerning our contribution are outlined below:

- We propose a multi-view SSL algorithm that handles efficiently both labeled (L) and unlabeled (U) data in the case of binary output variables.
- Instead of demanding two sufficient and redundant views, a random feature split is applied, thereby increasing the applicability and improving the performance of the finally formatted algorithm [38].
- We introduce a simple mechanism concerning the cardinality of unlabeled examples per different class that is mined for avoiding overfitting phenomena in cases where imbalanced datasets must be assessed.
- We insert a preprocess stage, where a pool of single learners is mined by a Static Ensemble Selection algorithm to extract a powerful soft-Voting ensemble per different classification problem, seeking to produce a more accurate and robust semi-supervised algorithm operating under small labeled ratio values.

Let the whole dataset (X) consist of n instances and k features, apart from the class variable (Y) that, in the context of this work, is restricted to be a binary one. Thus, without loss of generality, we assume that $y_i \in \{0,1\}$ for each labeled instance $\{l_i, 1 \leq i \leq n_l\}$, while each unlabeled instance $\{u_i, 1 \leq i \leq n_u\}$ is characterized by the absence of the corresponding y_i value. The parameters n_l and n_u represent the cardinality of L and U subsets, respectively. After having removed all missing values—leading to a new cardinality of total instances (n')—it is evident that the following equation holds:

$$n' = n_l + n_u \tag{1}$$

Besides holding both numeric and categorical features, all the features of the latter form are converted into binary ones, increasing the initial number of k features into k', in case X contains at least one of them. Otherwise, since no augmentation of the initial features has been applied, the next two quantities coincide: $k \equiv k'$. Under this generic approach, classification algorithms that cannot handle categorical data are not rejected by the total proposed process.

This choice seems safe enough, since it does not reject the adoption of any learner—this mainly refers to learning algorithms that cannot handle efficiently the existence of both numerical and categorical data—although the manipulation of heterogeneous features is an open issue [39]. Afterwards, without introducing any specific assumption about the relationship or the origination of any included feature, the available feature vector F: <$f_1, f_2, \ldots, f_{k'}$> is split into two newly formatted subsets F_1 and F_2, where $F = F_1 \cup F_2$. Hence, two different datasets X_1, X_2 are generated, respectively, both including disjoint feature sets, but sharing the same class variable Y. Therefore, the final hypothesis space could be summarized as follows: $F_{view}: X_{view} \to [0, 1]$, where view = 1, 2.

Through the above described methodology, the following two choices are enabled: either to apply a common learning strategy for both views, such as adopting the same learner, or tackling each view separately, depending on underlying properties, such as the views' cardinalities, independence or correlation assumptions that affect the views' internal structure or other kind of relationships that specify the nature of each view, since two distinct tasks have been raised. Following the majority of the existing approaches found in the literature and taking into consideration that a random feature split operates as an agnostic factor regarding the structure of the constructed views, the first approach was adopted in the present study [40].

Under this strategy, and before the common base learner is built per view, a preprocess stage is inserted. This aims to measure the rate of the imbalanced instances found in the provided training set and to define the number of the instances that have to be mined from each class per iteration (Mined$_{class0}$, Mined$_{class1}$). Due to the SSL concept, the quota of L and U subsets is defined by a labeled ratio value (R). Given this setting, the amount of the initial training set (L_{view}) is computed according to the following formula:

$$\text{InitSize} = R \times \text{size}(X), \forall \text{view} = 1, 2 \qquad (2)$$

The cardinalities of both classes are then computed (C_{max}, C_{min}) regarding the available L^0_{view}. The minimum of them is set equal to 1 (Mined$_{class0}$), while the other one is equal to $\lfloor C_{max}/C_{min} \rfloor$ (Mined$_{class1}$). In this way, the provided class distribution of the labeled instances is assumed to be representative of the total problem defined also by the unknown instances that must be assessed. Finally, these two variables are exploited during the learning stage to retrieve a suitable number of unlabeled instances per class during each iteration.

Now, as it regards the choice of the base learners, we selected five representative algorithms from different learning families, capturing a wide spectrum of properties, concerning both assets and defects, which should be combined and avoided, respectively, in order to construct appropriately an accurate and robust enough ensemble learner per dataset so as to initialize the co-training process [41]. For this purpose, our pool of classifiers (C) consists of support vector machines (SVMs) [42], k-nearest-neighbors (kNN) [43], a simple tree inducer (DT) from family of decision trees [44], naive Bayes (NB) [45], and logistic regression (LR) [46]. In order to keep the computational needs of the exported ensemble, we restrict the cardinality of classifier participants under our Voting scheme, setting this number equal to 2. Thus, we had to employ a soft variant of Voting scheme which takes into account the class-probabilities of each algorithm and combines these decisions through averaging process, instead of hard voting through on-off decisions [29], where the occurrence of ties with the even number of base learners would appear too frequent. Furthermore, the stage of averaging the decisions of each individual participant generally leads to the reduction of the ensemble's variance and helps to surpass the structure sensitivity that is usually detected in more unstable methods, considering the input data

To be more specific, if we assume that we tackle with a binary classification problem containing a set of labels $Y = \{0, 1\}$ and a feature space $X \in \mathbb{R}^k$, such that for any probabilistic classifier F holds the next function: $F: X \to Y$, then for each instance m the decision profile of learner j is a pair of class probabilities $[P_{j0}, P_{j1}]$ which sum up to 1. Consequently, the mechanism of a simple, without weighting factors, soft-Voting classifier, given an instance x_m, combines the decisions of all the candidate classification algorithms searching the most probable class (w) as follows:

$$\hat{y}_m = \arg\max_\omega \sum_{j=1}^{|p|} P_j(=\omega|x_m), \ y_m \in Y, \ m \in \{1,2,\ldots,n'\}, \quad (3)$$

The class with the largest average probability is exported as the prevalent one through this pipeline, where $\hat{y}_m \in Y$ and the notation of $|p|$ depicts the number of the combined classifiers.

Trying to uncover the function of our preprocess stage which constructs the base learner of the proposed co-training scheme, we had to refer that the ambition of any Static Ensemble Selection strategy is to construct a subset C^*, such that $C^* \subset C$ and $|C^*| = 2$, which satisfies better the chosen criteria for obtaining the most desired performance over test instances. In our case, we investigate the most compatible pair of learners that maximizes our proposed criterion under an unweighted soft-Voting scheme. Through this, we measure the number of instances for which the decision of the soft-Voting scheme remains correct when the two candidate participants disagree ($q_{corrected}$), normalized by the total amount of disagreements based on the label of the examined instances ($q_{disaggre}$), as well as the rate of non-common errors ($q_{common\ errors}/v$). To this end, we introduce the objective function of Equation (4), which is defined as a linear combination of the mentioned quantities:

$$Q^a_{soft}(i,j) = a * \frac{q^{i,j}_{corrected}}{q^{i,j}_{disagree}} + (1-a) * (1 - \frac{q^{i,j}_{common\ errors}}{v}), \quad (4)$$

$$0 \le a \le 1, \ i,j \in \{0,1,2\ldots,|C^*|\} \text{ with } i \neq j,$$

where a is a parameter to balance the importance between the included terms. Actually, the first one rewards the pair of classifiers that managed to act complementary, since the more times the confidence of the classifier that guessed correctly the corresponding class label overpowered against the erroneous one, the larger values this term records. On the other hand, the second term penalizes the pair of classifiers whose common decisions coincide with mislabeling cases by reducing its value when such behavior occurs. The parameter v symbolizes the cardinality of the validation set over which the rest of quantities are calculated. Giacinto and Roli called this diversity measure as "the double-fault measure" [47].

Although an analysis of the selected a value could raise the interest of further research, we selected the value of 0.5 for equal importance. Thus, for each examined dataset D, which contains both labeled and unlabeled data, we split the labeled set into train and validation set, in a same manner as the default k-fold-cross-validation strategy, applying the previously referred Static Ensemble Selection strategy so as to detect the most favorable pair of classifiers for our soft-Voting ensemble learner. In case that $q_{disaggre} = 0$, then a is set equal to 0, holding only the second term.

Exploiting the exported soft-Voting ensemble learner as the base learner of our co-training variant, each L^0_{view} is fitted with $Co(Votesoft(C^*_i, C^*_j)) \equiv Co(Vote^{SEC}_{soft})$—we use the notation C^*_i and C^*_j for the selected learners which are included into C^*—and the corresponding class probabilities for each unlabeled instance per view (u^i_{view}) are computed per iteration. Next, only the top-class0 and top-class1 instances per class are selected, based on the estimated confidence measure. Subsequently, these instances are exported by the current U subset (since both views share the common unlabeled set, it does not need to use the view index when referring to the U subset). Then, they are added to the training set of the opposite view along with the most prominent class label based on base learner's decision. Therefore, if the target variable of the m-th instance of U is categorized as class0 by the F_1 classifier (x_m: <$f_1, f_2, \ldots, f_{k'/2}$ with probclass0$_{first}$ > 0.5), then the L^{iter}_2 subset during the iter-th iteration has to be augmented with the same instance, using the corresponding features of the second view and the estimated class variable (x_m: <$f_{k'/2+1}, f_{k'/2+2}, \ldots, f_{k'}$|class0>).

According to this learning scheme whose main ambition is to teach two different learners of the same classification algorithm through mutual disagreement concept, each learner injects into the other the information that is retrieved by the supplied view per iteration. A more theoretical analysis of the error bounds that can be achieved through the disagreement-based concept in case of Co-training could be found in [48]. Since our strategy of constructing the base learner of co-training through a static ensemble selection mechanism $Vote_{soft}^{SEC}$, we assume that we provide an accurate enough algorithm whose both competence's performance and diversity's behavior have been verified through a validation set so as to avoid overfitting phenomena or heavy mislabeling learning behaviors.

To sum up, the pseudo-code of the introduced SEC strategy (SSoftEC) as well as the proposed co-training variant are presented in Algorithms 1 and 2, respectively.

Algorithm 1. SSoftEC *strategy*

Input:
L—labeled set
f—number of folds to split the L
C—pool of classification algorithms exporting class probabilities
α—value of balancing parameter
Main Procedure:
For each $i, j \in \{0, 1, \ldots, |C|\}$ and $i \neq j$ do
Set $iter = 0$, $Q_{soft}^a(i, j) = 0$
Split L to f separate folds: $\{L^{(1)}, L^{(2)}, \ldots, L^{(f)}\}$
While $iter \leq f$ do
    Train C_i, C_j on $L \setminus L^{(iter)}$
    Apply C_i, C_j on $L^{(iter)}$
    Update $Q_{soft}^a(i, j)$ according to Equation (4)
    $iter = iter + 1$
Output:
Return pair of indices i, j such that: $(i, j)^* : arg\ \max_{i,j} Q_{soft}^a(i, j)$.

Algorithm 2. *Ensemble based co-training variant*

Mode:
Pool-based scenario over a provided dataset $D = X_{n \times k} \cup Y_{n \times 1}$
x_i—vector with k features $<f_1, f_2, \ldots f_k> \forall\ 1 \leq i \leq n$
y_i—scalar class variable with $y_i \in \{0, 1\} \forall\ 1 \leq i \leq n$
$\{x_i, y_i\}$—i-th labeled instance (l^i) with $1 \leq i \leq n_l$
$\{x_i\}$—i-th unlabeled instance (u^i) with $1 \leq i \leq n_u$
F_{view}—separate feature sets with view $\in [1,2]$
$learner_{view}$—build of selected learner on corresponding View, $\forall view = 1, 2$
Input:
L^{iter}—labeled instances during iter-th iteration, $L^{iter} \subset D$
U^{iter}—unlabeled instances during iter-th iteration, $U^{iter} \subset D$
iter—number of combined executed iterations
MaxIter—maximum number of iterations
C—pool of classifiers $\equiv$ {SVM, kNN, DT, NB, LR}
(f, α)—number of folds to split the validation set during SEC and value of Equation (4)
Preprocess:
k'—number of features after having converted each categorical feature into binary
n'—number of instances after having removed instances with at least one missing value
C_j—instance cardinalities of both existing classes with $j \in$ {min, max}
$Mined_c$—define number of mined instances per class, where $c \in \{class_0, class_1\}$

Main Procedure:
Apply SSoftEC(L^0, f, C, α) and obtain C_i^*, C_j^*
Construct $Vote_{soft}(C_i^*, C_j^*)$
Set $iter = 0$
While $iter <$ MaxIter **do**
    For each $view$
        Train learner$_{view}$ on L_{view}^{iter}
        Assign class probabilities for each $u_i \in U^{iter}$
        For each $class$
            Detect the top Mined$_{class} \equiv$ Ind$_{view}$
            Update:
$$L_{view}^{iter+1} \leftarrow L_{view}^{iter} \cup \left\{x_j, \arg\max_{class} P(Y = class | X_{view}) \; \forall \; j \in Ind_{\sim view}\right\}$$
            (The sign $\sim$ view means the opposite view from the current.
$$U_{view}^{iter+1} \leftarrow U_{view}^{iter} \setminus \{x_j\} \forall j \in Ind_{\sim view}$$
$iter = iter + 1$
Output:
Use $Vote_{soft}(C_i^*, C_j^*)$ trained on L^{MaxIter} to predict class labels of test data.

4. Experimental Procedure and Results

For the purpose of our study a number of experiments were carried out using 27 benchmark datasets from UCI Machine Learning Repository [49] regarding binary classification problems (Table 1), where the sign # depicts the cardinality of the corresponding quantity. Note that the columns entitled # Features in Table 1, counts all the features apart from the class variable. These datasets have been partitioned into 10 equal-sized folds using the stratified 10-fold-CV resampling procedure so that each fold should have the same distribution as the entire dataset [50]. This process was repeated 10 times until all folds were used as the testing set and the results were averaged. Moreover, each fold was divided into two subsets, one labeled and the other one unlabeled, in accordance with a selected labeled ratio value (R) which is defined as follows:

$$R = |L_D|/(|L_D|+|U_D|). \quad (5)$$

Table 1. Description of datasets used from the UCI repository.

Dataset	# Instances	# Features	Dataset	# Instances	# Features
bands	365	19	monk-2	432	6
breast	277	48	pima	768	8
bupa	345	6	saheart	468	9
chess	3196	38	sick	3772	33
colic.orig	368	471	tic-tac-toe	958	27
diabetes	768	8	vote	435	16
heart-statlog	270	13	wdbc	569	30
kr-vs-kp	3196	40	wisconsin	683	9
mammographic	830	5			

#: the cardinality of the corresponding quantity.

In order to study the influence of the amount of labeled data in the training set, three different ratios were used, and in particular: 10%, 20%, and 30%. In general, the R (%) values over which researchers are interested are the smaller ones (R < 50%), so as to be consistent with the practical aspect of SSL scenario. The effectiveness of the proposed co-training scheme was compared to several co-training and self-training variants. For verifying the supremacy of the $Vote_{soft}^{SEC}$ as base classifier, we built the soft-Voting versions based on all pairs of the inserted pool of classifiers (C). Furthermore, the version that exploits the decisions of all the participants of C pool was implemented, as well as the individual variants without voting. Thus, 16 different supervised classifiers were exhibited as base learners, all imported by the scikit-learn Python library [51] and in particular:

The SVMs using Radial Basis Function as kernel inside its implementation, representing one universal learner that tries to separate instances using hyper-planes and 'Kernel-trick' [52],

- The k-Nearest Neighbor (kNN) instance-based learner [53] with k equal to 5, a very effective method for classification problems, using the Euclidean metric as a similarity measure to determine the distance between two instances,
- A simple Decision Tree (DT) algorithm, a variant of tree induction algorithms with large depth that split the feature space using 'gini' criterion [44],
- The NB probabilistic classifier, a simple and quite efficient classification algorithm based on the assumption that features are independent of each other given the class label [54],
- The Logistic Regression (LR), a well-known discriminative algorithm that assumes the log likelihood ratio of class distributions is linear in the provided examples. Its main function supports the binomial case of the target variable, exporting posterior probabilities in a direct way. In our implementation, L2-norm during penalization stage was chosen [55].

For simplicity, we made use of the following notation in the experiments, while the parameters' configuration for all applied classification methods is presented in Table 2:

- $C \equiv \{SVM, kNN, DT, NB, LR\}$, the list of participant classification algorithms,
- $Self(learner)$, where $learner \in C$,
- $Self(Vote(learner_i, learner_j))$, where $learner_i, learner_j \in C$ with $i \neq j$,
- $Self(Vote(all))$, where all participants of C are exploited under the Voting scheme,
- $Co(learner)$, where this kind of approach corresponds to the case that learner1 $\equiv$ learner2 $\equiv$ learner, with $learner \in C$,
- $Co(Vote(learner_i, learner_j))$, where $learner_i, learner_j \in C$ with $i \neq j$, and the ensemble Voting learner is the same for both views, similar with the previous scenario,
- $Co(Vote(all))$, where all participants of C are exploited under the Voting scheme for each view, and finally,
- $Co(Vote_{soft}^{SEC})$, which coincides with the proposed semi-supervised algorithm.

Table 2. Configuration of exploited algorithms' parameters.

Algorithm	Parameters
k-NN	Number of neighbors: 5
	Distance function: Euclidean distance
SVM	Kernel function: RBF
DT	Splitting criterion: gini
	Min instances per leaf = 2
LR	Norm: L2
NB	Gaussian distribution
Self-training	MaxIter = 20
Co-training	MaxIter = 20

As mentioned before, there are 10 different pairs of algorithms that can be formatted with a pool of five candidate classifiers. In addition, the case of applying each one individually takes also place, as well as the case that all participants of pool C are exploited under the same Voting stage. Thus, 16 self-training variants and 16 co-training variants are examined against the proposed co-training algorithm, which selects through a static selection strategy the soft-Voting ensemble base learner into its operation per different task. As it concerns the parameter f, it has been set equal to 10, leading to a 10-fold-cross-validation procedure per examined dataset. The next tables depict only one out of three different labeled ratio scenarios concerning the top five algorithms, based on total Friedman Ranking statistical process along with a smaller statistical comparison concerning only the top five algorithms. For a deeper analysis, the total results can be found in http://mL.math.upatras.gr/wp-content/uploads/2019/12/Official_results_co_training_ssoftec_voting.7z. Moreover, a pie chart has been provided in Figure 1, depicting the participation, into per centage style, of the pair of classifiers that were employed into the proposed strategy as base learner during all the experiments.

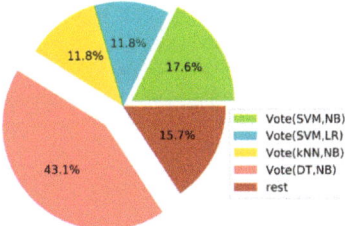

Figure 1. Pie chart depicting the participation of each combination inside our Static Ensemble strategy.

For evaluating the predictive performance of the proposed algorithm, three representative and widely used evaluation measures were adopted for measuring the obtained performance over the test set: classification accuracy, F_1-score, and Area Under the ROC Curve (AUC). Accuracy corresponds to the percentage of correctly classified instances, while F_1-score is an appropriate metric for imbalanced datasets and is defined as the harmonic mean of recall (r) and precision (p). In the case of a binary classification problem, they are defined as:

$$Accuracy = (tp + tn)/n \qquad (6)$$

$$F_1 score = 2 \times tp/(2 \times tp + fp + fn). \qquad (7)$$

where tp, tn, fp, fn, and n correspond to the number of true positive, true negative, false positive, false negative, and total number of instances, respectively. Finally, the latter one is related to the quality of the examined classifier ranking of any randomly chosen instance and is computed by aggregating the corresponding performance across all possible classification thresholds. The most favorable manner to visualize this metric is through plots of TPR vs. FPR or Sensitivity vs. (1-Specificity) relationship at different classification thresholds, where TPR stands for True Positive Rate, while FPR stands for False Positive Rate. Their analytical formulas are provided here:

$$TPR = tp/(tp + fn), \qquad (8)$$

$$FPR = fp/(fp + tn). \qquad (9)$$

The experimental results using 10% labeled ratio are summarized in Tables 3–5, where the best value per dataset is bold highlighted. Overall, it appears that the co-training Vote performs better than the corresponding self-training variants. Moreover, among the co-training variants employed, the proposed algorithm takes precedence over the rest on most of the datasets. In addition, we applied a familiar statistical tool to confirm the observed results. Hence, the Friedman Aligned Ranks [56] non-parametric test (significance level $\alpha = 0.05$) was used to compare all the employed SSL methods (Table 6). According to the calculated results, the algorithms are sorted from the best performer (lowest ranking) to the worst one (higher ranking). Therefore, it is statistically confirmed the supremacy of the $Co(Vote_{soft}^{SEC})$ algorithm, while the null hypothesis H_0 (i.e., the means of the results of two or more algorithms are the same) is rejected. Furthermore, the Nemenyi post-hoc test [57] ($\alpha = 0.05$) was applied to detect the specific differences between the algorithms, which is a commonly used non-parametric test for pairwise multiple comparisons. Table 6 includes the computed Critical

Difference (CD) which is the same for all the cases of this R-based scenario (CD = 2.27). It is statistically confirmed that the difference between the $Co(Vote_{soft}^{SEC})$ algorithm and the majority of the other methods is statistically significant in all examined metrics, thus verifying the predominance of the proposed co-training scheme. The fact also that the proposed algorithm outperforms the Vote (all) variants means that the implemented time-efficient SEC strategy provides a more accurate base learner for the field of SSL. Towards this direction, we visualize the performance of the proposed algorithm against $Co(Vote(all))$ for the examined metrics and the case of R = 90% via a violin plot which favors the comparison of the distribution of the achieved values per algorithm, including also some important statistical quantities: median, interquartile range, and 1.5× interquartile range (Figure 2). Therefore, we can deduce experimentally the success of the proposed approach, especially when generic binary datasets constitute the main issue to be tackled when the collected labeled instances are highly numerically restricted.

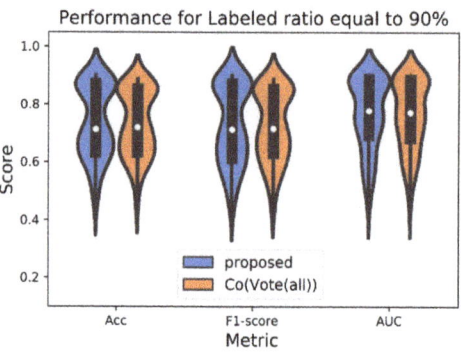

Figure 2. Violin plots of the proposed algorithm against $Co(Vote(all))$ approach over the three examined metrics.

Table 3. Classification accuracy (±stdev) values for the best five variants (labeled ratio 10%).

Dataset	Algorithms				
	$Co(Vote_{soft}^{SEC})$	$Co(Vote(all))$	$Co(Self(all))$	$Co(LR)$	$Co(Vote(DT,LR))$
bands	**0.668 ± 0.082**	0.668 ± 0.051	0.649 ± 0.036	0.627 ± 0.036	0.657 ± 0.084
breast	0.679 ± 0.065	0.696 ± 0.056	**0.7 ± 0.038**	0.693 ± 0.056	0.686 ± 0.114
bupa	0.571 ± 0.052	0.603 ± 0.085	0.529 ± 0.041	0.537 ± 0.042	**0.591 ± 0.057**
chess	**0.966 ± 0.013**	0.948 ± 0.007	0.955 ± 0.008	0.947 ± 0.011	0.962 ± 0.01
colic.ORIG	0.759 ± 0.055	0.757 ± 0.044	0.724 ± 0.077	0.73 ± 0.053	**0.768 ± 0.036**
diabetes	0.773 ± 0.038	**0.777 ± 0.032**	0.738 ± 0.048	0.765 ± 0.03	0.681 ± 0.059
h-statlog	0.719 ± 0.036	0.7 ± 0.027	0.707 ± 0.051	**0.77 ± 0.034**	0.656 ± 0.08
kr-vs-kp	**0.966 ± 0.019**	0.947 ± 0.008	0.953 ± 0.008	0.946 ± 0.012	0.963 ± 0.012
mammographic	0.788 ± 0.019	**0.796 ± 0.023**	0.793 ± 0.024	0.787 ± 0.025	0.78 ± 0.039
monk-2	**1 ± 0**	0.918 ± 0.019	0.88 ± 0.047	0.784 ± 0.029	**1 ± 0**
pima	0.691 ± 0.02	0.683 ± 0.025	0.686 ± 0.038	**0.697 ± 0.032**	0.652 ± 0.033
saheart	0.715 ± 0.021	0.713 ± 0.067	0.702 ± 0.036	**0.728 ± 0.02**	0.672 ± 0.038
sick	0.976 ± 0.003	0.967 ± 0.005	0.963 ± 0.006	0.95 ± 0.004	**0.978 ± 0.004**
tic-tac-toe	0.804 ± 0.07	0.805 ± 0.038	0.735 ± 0.037	**0.824 ± 0.036**	0.796 ± 0.057
vote	0.914 ± 0.026	0.891 ± 0.026	**0.916 ± 0.024**	0.889 ± 0.02	0.889 ± 0.033
wdbc	0.958 ± 0.021	0.963 ± 0.01	**0.972 ± 0.012**	0.956 ± 0.021	0.942 ± 0.031
wisconsin	0.974 ± 0.006	**0.98 ± 0.01**	0.972 ± 0.011	0.97 ± 0.016	0.936 ± 0.039

Bold highlighted means the best value per dataset.

Table 4. F_1-score (±stdev) values for the best five variants (labeled ratio 10%).

Dataset	Algorithms				
	$Co(Vote_{soft}^{SEC})$	$Co(Vote(all))$	$Co(Vote(DT,LR))$	$Co(Vote(all))$	$Co(LR)$
bands	**0.646 ± 0.096**	0.632 ± 0.068	0.643 ± 0.082	0.603 ± 0.049	0.594 ± 0.028
breast	0.652 ± 0.07	0.664 ± 0.076	**0.67 ± 0.118**	0.669 ± 0.032	0.649 ± 0.072
bupa	0.554 ± 0.049	**0.588 ± 0.085**	0.586 ± 0.059	0.502 ± 0.042	0.512 ± 0.032
chess	**0.966 ± 0.013**	0.947 ± 0.007	0.962 ± 0.01	0.955 ± 0.008	0.947 ± 0.011
colic.ORIG	0.745 ± 0.056	0.761 ± 0.043	**0.762 ± 0.039**	0.708 ± 0.083	0.735 ± 0.049
diabetes	0.765 ± 0.05	**0.769 ± 0.034**	0.68 ± 0.054	0.723 ± 0.055	0.75 ± 0.042
h-statlog	0.717 ± 0.036	0.7 ± 0.027	0.653 ± 0.082	0.706 ± 0.052	**0.77 ± 0.034**
kr-vs-kp	**0.966 ± 0.019**	0.947 ± 0.008	0.963 ± 0.012	0.953 ± 0.008	0.946 ± 0.012
mammographic	0.786 ± 0.019	**0.795 ± 0.022**	0.779 ± 0.039	0.791 ± 0.025	0.785 ± 0.026
monk-2	**1 ± 0**	0.918 ± 0.019	**1 ± 0**	0.879 ± 0.047	0.78 ± 0.03
pima	0.673 ± 0.018	0.664 ± 0.027	0.652 ± 0.034	0.664 ± 0.033	**0.685 ± 0.03**
saheart	0.65 ± 0.034	0.7 ± 0.073	0.675 ± 0.038	0.688 ± 0.039	**0.716 ± 0.019**
sick	0.974 ± 0.003	0.962 ± 0.007	**0.976 ± 0.004**	0.959 ± 0.008	0.939 ± 0.006
tic-tac-toe	0.807 ± 0.069	0.805 ± 0.037	0.799 ± 0.056	0.733 ± 0.038	**0.822 ± 0.036**
vote	0.914 ± 0.026	0.891 ± 0.026	0.888 ± 0.033	**0.916 ± 0.024**	0.888 ± 0.02
wdbc	0.958 ± 0.021	0.963 ± 0.01	0.943 ± 0.031	**0.972 ± 0.012**	0.956 ± 0.02
wisconsin	0.974 ± 0.006	**0.98 ± 0.01**	0.935 ± 0.04	0.972 ± 0.011	0.969 ± 0.016

Bold highlighted means the best value per dataset.

Table 5. Classification accuracy (±stdev) values for the best five variants (labeled ratio 10%).

Dataset	Algorithms				
	$Co(Vote_{soft}^{SEC})$	$Co(Vote(DT,LR))$	$Co(Vote(all))$	$Co(Vote(5NN,LR))$	$Self(Vote(all))$
bands	0.716 ± 0.083	0.652 ± 0.079	**0.764 ± 0.063**	0.716 ± 0.052	0.69 ± 0.06
breast	**0.701 ± 0.077**	0.671 ± 0.105	0.652 ± 0.062	0.661 ± 0.071	0.621 ± 0.081
bupa	0.584 ± 0.09	0.567 ± 0.065	**0.602 ± 0.08**	**0.602 ± 0.08**	0.54 ± 0.082
chess	0.99 ± 0.004	**0.992 ± 0.004**	0.985 ± 0.007	0.981 ± 0.006	**0.992 ± 0.005**
colic.ORIG	**0.835 ± 0.044**	0.817 ± 0.029	0.722 ± 0.079	0.813 ± 0.026	0.762 ± 0.097
diabetes	0.835 ± 0.037	0.772 ± 0.028	**0.854 ± 0.011**	0.824 ± 0.035	0.801 ± 0.039
h-statlog	0.765 ± 0.05	0.779 ± 0.053	**0.802 ± 0.024**	0.73 ± 0.051	0.769 ± 0.033
kr-vs-kp	0.99 ± 0.004	0.991 ± 0.004	0.985 ± 0.007	0.979 ± 0.006	**0.992 ± 0.005**
mammographic	0.857 ± 0.016	0.876 ± 0.021	0.856 ± 0.015	0.87 ± 0.021	0.874 ± 0.016
monk-2	0.996 ± 0.005	**1 ± 0**	**1 ± 0**	0.956 ± 0.018	0.958 ± 0.025
pima	0.655 ± 0.041	**0.681 ± 0.031**	0.68 ± 0.02	0.647 ± 0.041	0.651 ± 0.041
saheart	0.784 ± 0.057	0.764 ± 0.033	0.772 ± 0.019	**0.791 ± 0.028**	0.767 ± 0.046
sick	0.956 ± 0.011	0.961 ± 0.013	0.904 ± 0.024	0.917 ± 0.011	**0.963 ± 0.025**
tic-tac-toe	0.883 ± 0.031	0.903 ± 0.036	0.852 ± 0.054	**0.914 ± 0.035**	0.804 ± 0.041
vote	0.963 ± 0.005	**0.964 ± 0.006**	0.953 ± 0.018	0.958 ± 0.007	0.963 ± 0.007
wdbc	**0.995 ± 0.003**	0.992 ± 0.006	0.993 ± 0.004	**0.995 ± 0.003**	0.993 ± 0.006
wisconsin	**0.998 ± 0.002**	0.995 ± 0.007	0.996 ± 0.003	0.997 ± 0.003	0.997 ± 0.002

Bold highlighted means the best value per dataset.

Table 6. Friedman Rankings for all examined algorithms and statistical importance based on Nemenyi post-hoc test.

Friedman Ranking					
Acc		F_1-Score		AUC	
Algorithm	Rank	Algorithm	Rank	Algorithm	Rank
$Co(Vote_{soft}^{SEC})$	9.96	$Co(Vote_{soft}^{SEC})$	10.43	$Co(Vote_{soft}^{SEC})$	10.43
Co(Vote(all))	11.59	Co(Vote(all))	11.29	Co(Vote(DT,LR))	11.33
Self(Vote(all))	13.19	Co(Vote(DT,LR))	12.81	Co(Vote(all))	12.88
Co(LR)	13.36	Self(Vote(all))	13.28	Co(Vote(kNN,LR))	13.17
Co(Vote(DT,LR))	13.89	Co(LR)	13.34	Self(Vote(all))	13.66
Co(Vote(SVM,LR))	14.13	Co(DT)	13.44	Co(Vote(DT,GNB))	14.17
Co(DT)	14.41	Co(Vote(SVM,DT))	13.54	Self(Vote(DT,LR))	14.28
Co(Vote(SVM,DT))	14.58	Co(Vote(DT,GNB))	13.64	Co(Vote(SVM,LR))	14.98
Co(Vote(DT,GNB))	14.58	Self(Vote(DT,GNB))	14.31	Co(SVM)	14.98
Co(Vote(kNN,LR))	14.92	Co(Vote(kNN,DT))	14.54	Co(LR)	14.98
Self(Vote(DT,GNB))	15.33	Co(Vote(kNN,LR))	15.06	Self(Vote(DT,GNB))	15.24
Self(LR)	15.44	Self(DT)	15.12	Co(Vote(GNB,LR))	15.29
Co(Vote(kNN,DT))	15.64	Self(LR)	15.31	Self(LR)	16.01
Self(Vote(SVM,LR))	16.02	Self(Vote(kNN,DT))	15.46	Self(Vote(kNN,LR))	16.59
Self(DT)	16.04	Self(Vote(DT,LR))	15.78	Co(Vote(kNN,DT))	16.74
Self(Vote(kNN,DT))	16.61	Self(Vote(SVM,DT))	16.33	Co(Vote(kNN,GNB))	16.81
Self(Vote(DT,LR))	16.91	Co(Vote(SVM,LR))	16.33	Co(Vote(SVM,DT))	16.84
Self(Vote(kNN,LR))	16.96	Co(Vote(kNN,GNB))	17.05	Co(DT)	16.84
Co(Vote(SVM,kNN))	17.49	Self(Vote(kNN,LR))	17.21	Co(Vote(SVM,GNB))	17.06
Self(Vote(SVM,DT))	17.56	Self(Vote(SVM,LR))	17.38	Co(GNB)	17.06
Co(Vote(kNN,GNB))	18.06	Self(Vote(kNN,GNB))	17.69	Self(Vote(GNB,LR))	17.59
Co(kNN)	18.18	Co(kNN)	17.95	Self(Vote(SVM,LR))	18.09
Self(Vote(kNN,GNB))	18.34	Co(Vote(SVM,kNN))	19.20	Self(SVM)	18.09
Co(Vote(SVM,GNB))	19.45	Co(Vote(SVM,GNB))	19.59	Self(Vote(kNN,GNB))	18.42
Self(Vote(SVM,kNN))	19.46	Self(Vote(SVM,GNB))	19.83	Self(Vote(kNN,DT))	18.54
Self(Vote(SVM,GNB))	20.01	Self(Vote(GNB,LR))	19.88	Self(Vote(SVM,GNB))	18.62
Self(Vote(GNB,LR))	20.29	Co(Vote(GNB,LR))	20.02	Co(Vote(SVM,kNN))	19.41
Co(Vote(GNB,LR))	20.43	Self(kNN)	21.06	Co(kNN)	19.41
Co(SVM)	20.74	Self(GNB)	21.15	Self(GNB)	19.52
Self(kNN)	20.85	Self(Vote(SVM,kNN))	21.61	Self(Vote(SVM,DT))	21.03
Self(SVM)	21.89	Co(GNB)	21.84	Self(Vote(SVM,kNN))	22.62
Self(GNB)	22.14	Co(SVM)	24.21	Self(kNN)	24.67
Co(GNB)	22.56	Self(SVM)	25.33	Self(DT)	25.67

Adoption of more dedicated preprocessing stages oriented towards more specific problems should be applied, in order to boost the performance of the SSoftEC strategy and provide the Co-training scheme a more appropriate base learner [58–60]. However, in our generic experimental stage, which covers various applications, the proposed algorithm recorded a both robust and accurate enough performance, especially in the case of the F_1-score metric which is critical for real problems with class distribution different from the optimal, under a computational inexpensive manner, in contrast with DEC strategies that employ a new classifier search per test instance. The smoothing of the decisions that are produced through the proposed soft-Voting ensemble seems to favor the exported decision profile, since a large number of decisions that were initially misclassified based on individual predictions were reverted towards the ground truth label. While at the same time, numerous cases where the two participants disagree over the binary label were not affected. This happens because a large correct confidence value combined with a smaller incorrect one remains untouched under such a voting scheme, according to Equation (3).

5. Conclusions

In the present study, a soft-Voting ensemble-based co-training scheme through a Static Selection strategy operating under a random feature split, called $Co(Vote_{soft}^{SEC})$, was presented regarding binary classification problems. The proposed algorithm harnesses the benefits of the SSL approach and the ensemble methods that are built using heterogeneous approaches [61–63]. The experimental results using 27 benchmark datasets demonstrate the prevalence of the proposed algorithm in terms of classification accuracy and F_1-score, compared to several co-training and self-training variants, while using five different labeled ratios. Thus, the employment of a static selection classifier that tries to find a suitable combination among five provided classifiers for feeding appropriately an ensemble scheme seems that has favored the final predictive ability, without consuming much computational resources during the preprocess step per different task. This was successfully provoked by using soft-Voting strategy: for each class, the maximum averaged confidence is exported as the most prominent, taking into consideration the corresponding confidence values of both exploited learners.

Regarding our initial ambition, Co-training scheme seems more favorable for exploiting unlabeled instances and augmenting the initially collected labeled instances through the most reliable of the former, even during small labeled-ratio conditions, against Self-training approach. Furthermore, since the source and the structure of our examined datasets vary, application of the proposed method under more well-defined fields/tasks could be benefited by more advanced preprocessing stages, such as feature engineering, or tuning of participant learners, while specific criteria could be defined so as to avoid random split and converge to a more suitable feature split [40,64]. Increasing the cardinality and the diversification of the candidate classifiers should also be examined in following research, and especially the case when more strong classifiers are available, since their decision profile might demand a weighting soft or hard Voting scheme. In any case, the fact that unlabeled examples may boost the contribution of ensemble learners and their Selection strategies under SSL schemes seems to hold our assumption through our experimental stage [65]. Such strategies could also be used for selecting appropriate learners under more sophisticated ensemble structures like Stacking [66].

One interesting point, especially in the case that such a co-training algorithm should be applied to datasets that are characterized by more intense imbalanced classes, is the adoption of either more specified preprocess stages, such as the use of SMOTE algorithm, a well-known oversampling method that generates new instances or any of its descendants [67] or the embedding of similar methods inside the operation of base learner (s). Such an approach has been demonstrated recently in [68], where Rotation Forest, a popular ensemble learner based also on DTs, is combined with an under-sampling method so as to tackle this kind of issue. Regarding both the produced results and the fact that in semi-supervised scenarios the amount of collected data is much more restricted than the default supervised case, more sophisticated approaches for avoiding imbalanced datasets during the initialization of base learner, could be proven really promising for acquiring better learning rates [69]. Another interesting point is the expansion of the proposed scheme on the multi-class semi-supervised classification problem, such as in [70] where a new loss function was applied using gradient descent in functional space.

Moreover, appropriate experiments should be made towards the direction of recognizing the importance of the included features per view or among all the provided features, in case that random feature split is applied instead of merging all the distinct views into a compact but still heterogeneous view, so as to either propose a more detailed strategy for formatting the two separate views or applying feature reduction techniques that may favor the final predictive performance [64]. The "absent levels" problem [39] should also be studied under a SSL scenario, avoiding constructing views or preprocess feature sets that may lead to more implicit approaches, especially in real-life situations that the interpretability of the exported predictive model is of high priority for the business part or the corresponding field that is connected with the examined problem [71,72]. Construction of artificially generated data could be a safe strategy for excluding such conclusion. Otherwise, application to real-world data from totally different domains might infer biased decisions regarding the applicability to a wider range of datasets. Transfer learning, combined probably with Active Learning framework that permits the knowledge blending of human factors into the learning pipeline, could prove to be a valuable approach for exporting more robust classifiers by enriching the feature vector and/or applying more compatible modifications [73].

Finally, deep neural networks (DNNs) [74] could be employed under a co-training scheme to boost the predictive performance, fed with either raw data or other generic kinds of datasets. More specifically, long short term memory (LSTM) networks have already proven efficient enough when combined with SSL methods for constructing clinical support decision systems [75]. In case DNNs should be exploited, creation of new insights into inserted data could take place, providing either totally new view (s) or augmenting the existing one (s). Thus, several feature engineering approaches should be adopted to enhance the quality of the co-training scheme and possibly violate the assumption about the independent views less.

Author Contributions: Conceptualization, S.K. (Stamatis Karlos); methodology, S.K. (Stamatis Karlos), G.K. and S.K. (Sotiris Kotsiantis); software, S.K. (Stamatis Karlos); validation, S.K. (Stamatis Karlos), G.K. and S.K. (Sotiris Kotsiantis); formal analysis, G.K. and S.K. (Sotiris Kotsiantis); investigation, S.K. (Stamatis Karlos); resources, S.K. (Stamatis Karlos), G.K. and S.K. (Sotiris Kotsiantis); data curation, S.K. (Stamatis Karlos); writing—original draft preparation, S.K. (Stamatis Karlos), G.K.; writing—review and editing, S.K. (Stamatis Karlos), G.K. and S.K. (Sotiris Kotsiantis); visualization, S.K. (Stamatis Karlos); supervision, S.K. (Sotiris Kotsiantis); project administration, S.K. (Stamatis Karlos); funding acquisition, S.K. (Sotiris Kotsiantis). All authors have read and agreed to the published version of the manuscript.

Funding: This research received no external funding.

Conflicts of Interest: The authors declare no conflict of interest.

References

1. Schwenker, F.; Trentin, E. Pattern classification and clustering: A review of partially supervised learning approaches. *Pattern Recognit. Lett.* **2014**, *37*, 4–14. [CrossRef]
2. Kim, A.; Cho, S.-B. An ensemble semi-supervised learning method for predicting defaults in social lending. *Eng. Appl. Artif. Intell.* **2019**, *81*, 193–199. [CrossRef]
3. Li, J.; Wu, S.; Liu, C.; Yu, Z.; Wong, H.-S. Semi-Supervised Deep Coupled Ensemble Learning With Classification Landmark Exploration. *IEEE Trans. Image Process.* **2020**, *29*, 538–550. [CrossRef] [PubMed]
4. Kostopoulos, G.; Karlos, S.; Kotsiantis, S.; Ragos, O. Semi-supervised regression: A recent review. *J. Intell. Fuzzy Syst.* **2018**, *35*, 1483–1500. [CrossRef]
5. Alpaydin, E. *Introduction to Machine Learning*; MIT Press: Cambridge, MA, USA, 2010.
6. Ng, V.; Cardie, C. Weakly supervised natural language learning without redundant views. In Proceedings of the 2003 Human Language Technology Conference of the North American Chapter of the Association for Computational Linguistics, Edmonton, AB, Canada, 27 May–1 June 2003.
7. Blum, A.; Mitchell, T. Combining labeled and unlabeled data with co-training. In Proceedings of the Eleventh Annual Conference on Computational Learning Theory—COLT' 98, New York, NY, USA, 24–26 July 1998; pp. 92–100.
8. Zhou, Z.-H.; Li, M. Tri-training: Exploiting unlabeled data using three classifiers. *IEEE Trans. Knowl. Data Eng.* **2005**, *17*, 1529–1541. [CrossRef]
9. Zhu, X.; Goldberg, A.B. *Introduction to Semi-Supervised Learning*; Morgan & Claypool Publishers: Williston, VN, USA, 2009.
10. Zhou, Z.-H. *Ensemble Methods: Foundations and Algorithms*; Taylor & Francis: Abingdon, UK, 2012.
11. Zhou, Z.-H. When semi-supervised learning meets ensemble learning. *Front. Electr. Electron. Eng. China* **2011**, *6*, 6–16. [CrossRef]
12. Sinha, A.; Chen, H.; Danu, D.G.; Kirubarajan, T.; Farooq, M. Estimation and decision fusion: A survey. *Neurocomputing* **2008**, *71*, 2650–2656. [CrossRef]
13. Wu, Y.; He, J.; Man, Y.; Arribas, J.I. Neural network fusion strategies for identifying breast masses. In Proceedings of the 2004 IEEE International Joint Conference on Neural Networks (IEEE Cat. No. 04CH37541), Budapest, Hungary, 25–29 July 2004; pp. 2437–2442.

14. Wu, Y.; Arribas, J.I. Fusing output information in neural networks: Ensemble performs better. In Proceedings of the 25th Annual International Conference of the IEEE Engineering in Medicine and Biology Society (IEEE Cat. No. 03CH37439), Cancun, Mexico, 17–21 September 2003; pp. 2265–2268.
15. Livieris, I.; Kanavos, A.; Tampakas, V.; Pintelas, P. An auto-adjustable semi-supervised self-training algorithm. *Algorithms* **2018**, *11*, 139. [CrossRef]
16. Britto, A.S.; Sabourin, R.; Oliveira, L.E.S. Dynamic selection of classifiers—A comprehensive review. *Pattern Recognit.* **2014**, *47*, 3665–3680. [CrossRef]
17. Hou, C.; Xia, Y.; Xu, Z.; Sun, J. Semi-supervised learning competence of classifiers based on graph for dynamic classifier selection. In Proceedings of the IEEE 2016 23rd International Conference on Pattern Recognition (ICPR), Cancun, Mexico, 4–8 December 2016; pp. 3650–3654.
18. Jiang, Z.; Zhang, S.; Zeng, J. A hybrid generative/discriminative method for semi-supervised classification. *Knowl. Based Syst.* **2013**, *37*, 137–145. [CrossRef]
19. Ceci, M.; Pio, G.; Kuzmanovski, V.; Džeroski, S. Semi-supervised multi-view learning for gene network reconstruction. *PLoS ONE* **2015**, *10*, e0144031. [CrossRef] [PubMed]
20. Zhu, X.; Goldberg, A.B. Introduction to Semi-Supervised Learning. *Synth. Lect. Artif. Intell. Mach. Learn.* **2009**, *3*, 1–130. [CrossRef]
21. Nigam, K.; Ghani, R. Analyzing the effectiveness and applicability of co-training. In Proceedings of the Ninth International Conference on Information and Knowledge Management, New York, NY, USA, 6–11 November 2000; pp. 86–93. [CrossRef]
22. Yu, N. Exploring C o-training strategies for opinion detection. *J. Assoc. Inf. Sci. Technol.* **2014**, *65*, 2098–2110. [CrossRef]
23. Lin, W.-Y.; Lo, C.-F. Co-training and ensemble based duplicate detection in adverse drug event reporting systems. In Proceedings of the IEEE International Conference on Bioinformatics and Biomedicine, Shanghai, China, 18–21 December 2013; pp. 7–8.
24. Culp, M.; Michailidis, G. A co-training algorithm for multi-view data with applications in data fusion. *J. Chemom.* **2009**, *23*, 294–303. [CrossRef]
25. Wehrens, R.; Mevik, B.-H. The pls package: Principal component and partial least squares regression in R. *J. Stat. Softw.* **2007**, *18*, 1–23. [CrossRef]
26. Levatić, J.; Ceci, M.; Kocev, D.; Džeroski, S. Self-training for multi-target regression with tree ensembles. *Knowl. Based Syst.* **2017**, *123*, 41–60. [CrossRef]
27. Li, M.; Zhou, Z.-H. Improve Computer-Aided Diagnosis With Machine Learning Techniques Using Undiagnosed Samples. *IEEE Trans. Syst. Man Cybern. Part A Syst. Hum.* **2007**, *37*, 1088–1098. [CrossRef]
28. Deng, C.; Guo, M.Z. A new co-training-style random forest for computer aided diagnosis. *J. Intell. Inf. Syst.* **2011**, *36*, 253–281. [CrossRef]
29. Liu, C.; Yuen, P.C. A Boosted Co-Training Algorithm for Human Action Recognition. *IEEE Trans. Circuits Syst. Video Technol.* **2011**, *21*, 1203–1213. [CrossRef]
30. Kostopoulos, G.; Karlos, S.; Kotsiantis, S.B. Multi-view Learning for Early Prognosis of Academic Performance: A Case Study. *IEEE Trans. Learn. Technol.* **2019**, *12*, 212–224. [CrossRef]
31. Pio, G.; Malerba, D.; D'Elia, D.; Ceci, M. Integrating microRNA target predictions for the discovery of gene regulatory networks: A semi-supervised ensemble learning approach. *BMC Bioinform.* **2014**, *15*, S4. [CrossRef] [PubMed]
32. Dietterich, T.G. Ensemble Methods in Machine Learning. *Mult. Classif. Syst.* **2000**, *1857*, 1–15. [CrossRef]
33. Bolón-Canedo, V.; Alonso-Betanzos, A. *Recent Advances in Ensembles for Feature Selection*; Springer International Publishing: Cham, Switzerland, 2018.
34. Azizi, N.; Farah, N. From static to dynamic ensemble of classifiers selection: Application to Arabic handwritten recognition. *Int. J. Knowl. Based Intell. Eng. Syst.* **2012**, *16*, 279–288. [CrossRef]
35. Mousavi, R.; Eftekhari, M.; Rahdari, F. Omni-Ensemble Learning (OEL): Utilizing Over-Bagging, Static and Dynamic Ensemble Selection Approaches for Software Defect Prediction. *Int. J. Artif. Intell. Tools* **2018**, *27*, 1850024. [CrossRef]
36. Tsoumakas, G.; Partalas, I.; Vlahavas, I. An Ensemble Pruning Primer. *Appl. Supervised Unsupervised Ensemble Methods* **2009**, *245*, 1–13. [CrossRef]

37. Cruz, R.M.O.; Sabourin, R.; Cavalcanti, G.D.C. Analyzing different prototype selection techniques for dynamic classifier and ensemble selection. In Proceedings of the IEEE 2017 International Joint Conference on Neural Networks (IJCNN), Anchorage, AK, USA, 14–19 May 2017; pp. 3959–3966.
38. Zhao, J.; Xie, X.; Xu, X.; Sun, S. Multi-view learning overview: Recent progress and new challenges. *Inf. Fusion* **2017**, *38*, 43–54. [CrossRef]
39. Au, T.C. Random Forests, Decision Trees, and Categorical Predictors: The "Absent Levels" Problem. *J. Mach. Learn. Res.* **2018**, *19*, 1–30.
40. Ling, C.X.; Du, J.; Zhou, Z.-H. When does Co-training Work in Real Data? *Adv. Knowl. Discov. Data Min. Proc.* **2009**, *5476*, 596–603.
41. Ni, Q.; Zhang, L.; Li, L. A Heterogeneous Ensemble Approach for Activity Recognition with Integration of Change Point-Based Data Segmentation. *Appl. Sci.* **2018**, *8*, 1695. [CrossRef]
42. Platt, J.C. Probabilistic Outputs for Support Vector Machines and Comparisons to Regularized Likelihood Methods. *Adv. Large Margin Classif.* **1999**, *10*, 61–74.
43. Garcia, E.K.; Feldman, S.; Gupta, M.R.; Srivastava, S. Completely lazy learning. *IEEE Trans. Knowl. Data Eng.* **2010**, *22*, 1274–1285. [CrossRef]
44. Loh, W.-Y. Classification and regression trees. *Wiley Interdiscip. Rev. Data Min. Knowl. Discov.* **2011**, *1*, 14–23. [CrossRef]
45. Zheng, F.; Webb, G. A comparative study of semi-naive Bayes methods in classification learning. In Proceedings of the 4th Australas Data Mining Conference AusDM05 2005, Sydney, Australia, 5–6 December 2005; pp. 141–156.
46. Samworth, R.J. Optimal weighted nearest neighbour classifiers. *arXiv* **2011**, arXiv:1101.5783v3. [CrossRef]
47. Giacinto, G.; Roli, F. Design of effective neural network ensembles for image classification purposes. *Image Vis. Comput.* **2001**, *19*, 699–707. [CrossRef]
48. Wang, W.; Zhou, Z.-H. Theoretical Foundation of Co-Training and Disagreement-Based Algorithms. *arXiv* **2017**, arXiv:1708.04403.
49. Dua, D.; Graff, C. UCI Machine Learning Repository. Available online: http://archive.ics.uci.edu/ml (accessed on 1 November 2019).
50. James, G.; Witten, D.; Hastie, T.; Tibshirani, R. *An Introduction to Statistical Learning: with Applications in R*; Springer: New York, NY, USA, 2013.
51. Pedregosa, F.; Varoquaux, G.; Gramfort, A.; Michel, V.; Thirion, B.; Grisel, O.; Blondel, M.; Prettenhofer, P.; Weiss, R.; Dubourg, V.; et al. Scikit-learn: Machine learning in Python. *J. Mach. Learn. Res.* **2011**, *12*, 2825–2830.
52. Chang, C.; Lin, C. LIBSVM: A Library for Support Vector Machines. *ACM Trans. Intell. Syst. Technol.* **2011**, *2*, 1–39. [CrossRef]
53. Aha, D.W.; Kibler, D.; Albert, M.K. Instance-Based Learning Algorithms. *Mach. Learn.* **1991**, *6*, 37–66. [CrossRef]
54. Rish, I. An empirical study of the naive Bayes classifier. In Proceedings of the IJCAI 2001 Workshop on Empirical Methods in Artificial Intelligence, Seattle, WA, USA, 4–6 August 2001; pp. 41–46.
55. Sperandei, S. Understanding logistic regression analysis. *Biochem. Medica* **2014**, *24*, 12–18. [CrossRef]
56. Hodges, J.L.; Lehmann, E.L. Rank methods for combination of independent experiments in analysis of variance. *Ann. Math. Stat.* **1962**, *33*, 482–497. [CrossRef]
57. Hollander, M.; Wolfe, D.A.; Chicken, E. *Nonparametric Statistical Methods*; John Wiley & Sons, Inc.: Hoboken, NJ, USA, 2014.
58. Kumar, G.; Kumar, K. The Use of Artificial-Intelligence-Based Ensembles for Intrusion Detection: A Review. *Appl. Comput. Intell. Soft Comput.* **2012**, *2012*, 850160. [CrossRef]
59. Karlos, S.; Kaleris, K.; Fazakis, N. Optimized Active Learning Strategy for Audiovisual Speaker Recognition. In Proceedings of the 20th International Conference on Speech and Computer SPECOM 2018, Leipzig, Germany, 18–22 September 2018; pp. 281–290.
60. Tencer, L.; Reznakova, M.; Cheriet, M. Summit-Training: A hybrid Semi-Supervised technique and its application to classification tasks. *Appl. Soft Comput. J.* **2017**, *50*, 1–20. [CrossRef]
61. Tanha, J.; van Someren, M.; Afsarmanesh, H. Semi-supervised self-training for decision tree classifiers. *Int. J. Mach. Learn. Cybern.* **2017**, *8*, 355–370. [CrossRef]

62. Chapelle, O.; Schölkopf, B.; Zien, A. Metric-Based Approaches for Semi-Supervised Regression and Classification. In *Semi-Supervised Learning*; MIT Press: Cambridge, MA, USA, 2006; pp. 420–451.
63. Wainer, J. Comparison of 14 different families of classification algorithms on 115 binary datasets. *arXiv* **2016**, arXiv:1606.00930.
64. Yaslan, Y.; Cataltepe, Z. Co-training with relevant random subspaces. *Neurocomputing* **2010**, *73*, 1652–1661. [CrossRef]
65. Zhang, M.-L.; Zhou, Z.-H. Exploiting unlabeled data to enhance ensemble diversity. *Data Min. Knowl. Discov.* **2013**, *26*, 98–129. [CrossRef]
66. Karlos, S.; Fazakis, N.; Kotsiantis, S.; Sgarbas, K. Self-Trained Stacking Model for Semi-Supervised Learning. *Int. J. Artif. Intell. Tools* **2017**, *26*. [CrossRef]
67. Barua, S.; Islam, M.M.; Yao, X.; Murase, K. MWMOTE–Majority Weighted Minority Oversampling Technique for Imbalanced Data Set Learning. *IEEE Trans. Knowl. Data Eng.* **2014**, *26*, 405–425. [CrossRef]
68. Guo, H.; Diao, X.; Liu, H. Embedding Undersampling Rotation Forest for Imbalanced Problem. *Comput. Intell. Neurosci.* **2018**, *2018*, 6798042. [CrossRef]
69. Vluymans, S. Learning from Imbalanced Data. In *Dealing with Imbalanced and Weakly Labelled Data in Machine Learning Using Fuzzy and Rough Set Methods. Studies in Computational Intelligence*; Springer: Cham, Switzerland, 2019; pp. 81–110.
70. Tanha, J. MSSBoost: A new multiclass boosting to semi-supervised learning. *Neurocomputing* **2018**, *314*, 251–266. [CrossRef]
71. Chuang, C.-L. Application of hybrid case-based reasoning for enhanced performance in bankruptcy prediction. *Inf. Sci.* **2013**, *236*, 174–185. [CrossRef]
72. Ribeiro, M.T.; Singh, S.; Guestrin, C. "Why Should I Trust You?": Explaining the Predictions of Any Classifier. In Proceedings of the 22nd ACM SIGKDD International Conference on Knowledge Discovery and Data Mining—KDD'16, San Francisco, CA, USA, 13–17 August 2016; ACM Press: New York, NY, USA, 2016; pp. 1135–1144.
73. Kale, D.; Liu, Y. Accelerating Active Learning with Transfer Learning. In Proceedings of the 2013 IEEE 13th International Conference on Data Mining, Dallas, TX, USA, 7–10 December 2013; pp. 1085–1090. [CrossRef]
74. Nielsen, M.A. Neural Networks and Deep Learning. 2015. Available online: http://neuralnetworksanddeeplearning.com (accessed on 1 November 2019).
75. Chen, D.; Che, N.; Le, J.; Pan, Q. A co-training based entity recognition approach for cross-disease clinical documents. *Concurr. Comput. Pract. Exp.* **2018**, e4505. [CrossRef]

© 2020 by the authors. Licensee MDPI, Basel, Switzerland. This article is an open access article distributed under the terms and conditions of the Creative Commons Attribution (CC BY) license (http://creativecommons.org/licenses/by/4.0/).

Article

GeoAI: A Model-Agnostic Meta-Ensemble Zero-Shot Learning Method for Hyperspectral Image Analysis and Classification

Konstantinos Demertzis * and Lazaros Iliadis

Department of Civil Engineering, School of Engineering, Democritus University of Thrace, 67100 Xanthi, Greece; liliadis@civil.duth.gr
* Correspondence: kdemertz@fmenr.duth.gr; Tel.: +30-694-824-1881

Received: 13 February 2020; Accepted: 4 March 2020; Published: 7 March 2020

Abstract: Deep learning architectures are the most effective methods for analyzing and classifying Ultra-Spectral Images (USI). However, effective training of a Deep Learning (DL) gradient classifier aiming to achieve high classification accuracy, is extremely costly and time-consuming. It requires huge datasets with hundreds or thousands of labeled specimens from expert scientists. This research exploits the MAML++ algorithm in order to introduce the *Model-Agnostic Meta-Ensemble Zero-shot Learning* (MAME-ZsL) approach. The MAME-ZsL overcomes the above difficulties, and it can be used as a powerful model to perform Hyperspectral Image Analysis (HIA). It is a novel optimization-based Meta-Ensemble Learning architecture, following a *Zero-shot Learning* (ZsL) prototype. To the best of our knowledge it is introduced to the literature for the first time. It facilitates learning of specialized techniques for the extraction of user-mediated representations, in complex Deep Learning architectures. Moreover, it leverages the use of first and second-order derivatives as pre-training methods. It enhances learning of features which do not cause issues of exploding or diminishing gradients; thus, it avoids potential overfitting. Moreover, it significantly reduces computational cost and training time, and it offers an improved training stability, high generalization performance and remarkable classification accuracy.

Keywords: model-agnostic meta-learning; ensemble learning; GIS; hyperspectral images; deep learning; remote sensing; scene classification; geospatial data; Zero-shot Learning

1. Introduction

Hyperspectral image analysis and classification is a timely special field of Geoinformatics which has attracted much attention recently. This has led to the development of a wide variety of new approaches, exploiting both spatial and spectral content of images, in order to optimally classify them into discrete components related to specific standards. Typical information products obtained by the above approaches are related to diverse areas; namely: ground cover maps for environmental Remote Sensing; surface mineral maps used in geological applications; vegetation species maps, employed in agricultural-geoscience studies and in urban mapping. Recent developments in optical sensor technology and Geoinformatics (GINF), provide multispectral, Hyperspectral (HyS) and panchromatic images at very high spatial resolution. Accurate and effective HyS image analysis and classification is one of the key applications which can enable the development of new decision support systems. They can provide significant opportunities for business, science and engineering in particular. Automatic assignment of a specific semantic label to each object of a HyS image (according to its content) is one of the most difficult problems of GINF Remote Sensing (RES).

With the available HyS resolution, subtle objects and materials can be extracted by HyS imaging sensors with very narrow diagnostic spectral bands. This can be achieved for a variety of purposes, such as detection, urban planning, agriculture, identification, surveillance and quantification. HyS image analysis enables the characterization of objects of interest (e.g., land cover classes) with unprecedented accuracy, and keeps inventories up to date. Improvements in spectral resolution have called for advances in signal processing and exploitation algorithms.

A Hyperspectral image is a 3D data cube, which contains two-dimensional spatial information (image feature) and one-dimensional spectral information (spectral bands). Especially, the spectral bands occupy very fine wavelengths. Additionally, image features related to land cover and shape disclose the disparity and association among adjacent pixels from different directions at a confident wavelength. This is due to its vital applications in the design and management of soil resources, precision farming, complex ecosystem/habitat monitoring, biodiversity conservation, disaster logging, traffic control and urban mapping.

It is well known that increasing data dimensionality and high redundancy between features might cause problems during data analysis. There are many significant challenges which need to be addressed when performing HyS image classification. Primarily, supervised classification faces a challenge related to the imbalance between high dimensionality and incomplete accessibility of training samples, or to the presence of mixed pixels in the data. Further, it is desirable to integrate the essential spatial and spectral information, so as to combine the complementary features which stem from source images.

Deep Learning methodologies have significantly contributed towards the evolution and development of HyS image analysis and classification [1]. Deep Learning (DL) is a branch of computational intelligence which uses a series of algorithms that model high-level abstraction data using a multi-level processing architecture.

It is difficult for all Deep Learning algorithms to achieve satisfactory classification results with limited labeled samples, despite their undoubtedly well-established functions and their advantages. The approaches with the highest classification accuracy and generalization ability fall under the supervised learning umbrella. For this reason, especially in the case of Ultra-Spectral Images, huge datasets with hundreds or thousands of specimens labeled by experts are required [2]. This process is very expensive and time consuming.

In the case of supervised image classification, the input image is processed by a series of operations performed at different neuronal levels. Eventually, the output generates a probability distribution for all possible classes (usually using the *Softmax* function). Softmax is a function which takes an input vector Z of k real numbers and normalizes it into a probability distribution consisting of k probabilities, proportional to the exponentials of the input numbers [3].

$$\sigma(z)_j = \frac{e^{z_j}}{\sum_{k=1}^{K} e^{z_k}} \quad j = 1,\ldots,k \text{ where } \sigma : R^k \to R^k \; Z = (z_1,\ldots,z_k) \in R^k \qquad (1)$$

For example, if we try to classify an image as L_{im-a}, or L_{im-b}, or L_{im-c}, or L_{im-d}, then we generate four probabilities for each input image, indicating the respective probabilities of the image belonging to each of the four categories. There are two important points to be mentioned here. First, during the training process, we require a large number of images for each class (L_{im-a}, L_{im-b}, L_{im-c}, L_{im-d}). Secondly, if the network is only trained for the above four image classes, then we cannot expect to test it for any other class; e.g., "L_{im-x}." If we want our model to sort images as well, then we need to get many "L_{im-x}" images and to rebuild and retrain the model [3]. There are cases where we do not have enough data for each category, or the classes are huge but also dynamically changing. Thus, the cost of data collection and periodic retraining is enormous. A reliable solution should be sought in these cases. In contrast, k-shot learning is a framework within which the network is called upon to learn quickly and with few examples. During training, a limited number of examples from diverse classes with their labels are introduced. The network is required to learn general characteristics of the problem, such as features which are either common to the samples of the same class, or unique features which differentiate and eventually separate the classes.

In contrast to the learning process of the traditional neural networks, it is not sufficient for the network to learn good representations of the training classes, as the testing classes are distinct and they are not presented in training. However, it is desirable to learn features which distinguish the existing classes.

The evaluation process consists of two distinct stages of the following format [4]:

Step 1: Given k examples (value of k-shot), if $k = 1$, then the process is called *one-shot*; if $k = 5$, *five-shot*, and so on. The parameter k represents the number of labeled samples given to the algorithm by each class. By considering these samples, which comprise the support set, the network is required to classify and eventually adapt to existing classes.

Step 2: Unknown examples of the labeled classes are presented randomly, unlike the ones presented in the previous step, which the network is called to correctly classify. The set of examples in this stage is known as the query set.

The above procedure (steps) is repeated many times using random classes and examples which are sampled from the testing-evaluation set.

As it is immediately apparent from the description of the evaluation process, as the number of classes increases, the task becomes more difficult, because the network has to decide between several alternatives. This means that *Zero-shot Learning* [5] is clearly more difficult than the *one-shot*, which is more difficult than the *five-shot*, and so on. Although humans have the ability to cope with this process, traditional ANN require many more examples to generalize effectively, in order to achieve the same degree of performance. The limitation of these learning approaches is that the model has access to minimum samples from each class and the validation process is performed by calculating the cross-entropy error of the test set. Specifically, in the cases of one-shot and *Zero-shot Learning* (ZsL), only one example each of the candidate classes and only meta-data is shown at the evaluation stage.

Overall, *k-shot learning* is a perfect example of a problematic area, where specialized solutions are needed to design and train systems capable to learn very quickly from a small support set, containing only 1–5 samples per class. These systems can offer strong generalization to a corresponding target set. A successful exploitation of the above k-shot learning cases is provided by meta-learning techniques which can be used to deliver effective solutions [6].

In this work, we propose a new classification model, which is based on zero-shot philosophy, named MAME-ZsL. The significant advantages of the proposed algorithm is that it reduces computational cost and training time; it avoids potential overfitting by enhancing the learning of features which do not cause issues of exploding or diminishing gradients; and it offers an improved training stability, high generalization performance and remarkable classification accuracy. The superiority of the proposed model refers to the fact that the instances in the testing set belong to classes which were not contained in the training set. In contrast, the traditional supervised state-of-the-art Deep Learning models were trained with labeled instances from all classes. The performance of the proposed model was evaluated against state-of-the-art supervised Deep Learning models. The presented numerical experiments provide convincing arguments regarding the classification efficiency of the proposed model.

2. Meta-Learning

It is a field of machine learning where advanced learning algorithms are applied to the data and metadata of a given problem. The models "*learn to learn*" [7] from previous learning processes or previous sorting tasks they have completed [8]. It is an advanced form of learning where computational models, which usually consist of multiple levels of abstraction, can improve their learning ability. This is achieved by learning some or all of their own building blocks, through the experience gained in handling a large number of tasks. Their building blocks which are "*learning to learn*" can be optimizers, loss functions, initializations, Learning Rates, updated functions and architectures.

In general, for real physical modeling situations, the input patterns with and without tags are derived from the same boundary distribution or they follow a common cluster structure. Thus, classified data can contribute to the learning process, while correspondingly useful information related to the exploration of the data structure of the general set can be extracted from the non-classified data. This information can be combined with knowledge originating from prior learning processes or from completed prior classification tasks. Based on the above theory, *meta-learning* techniques can discover the structure of the data, by allowing new tasks to be learned quickly. This is achieved by using different types of metadata, such as the properties of the learning problem, the properties of the algorithm used (e.g., performance measures) or the patterns derived from data from a previous problem. This process employs knowledge from unknown cases sampled from real-world distribution of examples, aiming to enhance the outcome of the learning task. In this way it is possible to learn, select, change or combine different learning algorithms to effectively solve a given problem.

Meta-learning is achieved by conceptually dividing learning in two levels. The inner-most levels acquire specific knowledge for specific tasks (e.g., fine-tuning a model on a new dataset), while the outer-most levels acquire across-task knowledge (e.g., learning to transfer tasks more efficiently).

If the inner-most levels are using learnable parameters, outer-most optimization process can meta-learn the parameters of such components, thereby enabling automatic learning of inner-loop components.

A *meta-learning* system should combine the following three requirements [9,10]:

1. The system must include a learning sub-system.
2. Experience must be derived from the use of extracted knowledge from metadata related to the dataset under consideration, or from previous learning tasks completed in similar or different fields.
3. Learning biases should be dynamically selected.

Employing a generic approach, a credible meta-learning model should be trained in a variety of learning tasks and it should be optimized for the best performance in generalizing tasks, including potentially unknown ones. Each task is associated with a dataset D, containing attribute vectors and class labels in a supervised learning problem. The optimal parameters of the model are [11]:

$$\theta^* = \arg\min_\theta \mathbb{E}_{D \sim P(D)}[L_\theta(D)] \tag{2}$$

It looks similar to a normal learning process, but each data set is considered as a data sample.

The dataset D is divided in two parts, a training set S and a set of predictions B for validation and testing.

$$D = \langle S, B \rangle \tag{3}$$

The dataset D includes pairs of vectors and labels so which:

$$D = \{(x_i, y_i)\} \tag{4}$$

Each label belongs to a known label set L.

Let us consider a classifier f_θ. The parameter θ extracts the probability x, $P_\theta(y|x)$ of a data point to belong to class y, given by the attribute vector. Optimal parameters should maximize the likelihood of identifying true labels in multiple training batches $B \subset D$:

$$\theta^* = \arg\max_\theta \mathbb{E}_{(x,y) \in D}[P_\theta(y|x)] \tag{5}$$

$$\theta^* = \arg\max_\theta \mathbb{E}_{B \subset D}\left[\sum_{(x,y) \in B} P_\theta(y|x)\right] \tag{6}$$

The aim is to reduce the prediction error in data samples with unknown labels, in which there is a small set of "fast learning" support which can be used for "fine-tuning".

Fast learning is a trick which creates a "fake" dataset containing a small subset of labels (to avoid exposing all labels to the model). During the optimization process, various modifications take place, aiming to achieve rapid learning.

A brief step-by-step description of the whole process is presented below [11]:

1. Development of a subset of labels $L_s \subset L$
2. Development of a training subset $S^L \subset D$ and a forecast subset $B^L \subset D$. Both of them include data points with labels belonging to the subset L_s, $y \in L_s$, $\forall (x, y) \in S^L, B^L$.
3. The optimization procedure uses B^L to calculate the error and to update the model parameters via back propagation. This is done in the same way as it is used in a simple supervised learning model.

In this way each sample pair (S^L, B^L) can be considered to be a data point. The model is trained so that it can generalize to new unknown datasets.

The following function (Equation (7)) is a modification of the supervised learning model. The symbols of the meta-learning process have been added:

$$\theta^* = argmax_\theta \mathbb{E}_{L_s \subset L}\left[\mathbb{E}_{S^L \subset D, B^L \subset D}\left[\sum_{(x,y) \in B^L} P_\theta(x, y, S^L)\right]\right] \quad (7)$$

There are three *meta-learning* modeling approaches, as presented below [11] and the Table 1:

a. Model-based: These are techniques based on the use of circular networks with external or internal memory. They update their parameters quickly with minimal training steps. This can be achieved through their internal architecture or by using other control models. *Memory-augmented neural networks* and *meta networks* are characteristic cases of *model-based meta-learning* techniques.
b. Metrics-based: These are techniques based on learning effective distance measurements which can offer generalization. The core concept of their operation is similar to that of the nearest neighbors algorithms, where they aim to learn a measurement or a distance from objects. The concept of a good metric depends on the problem, as it should *represent* the relationship between inputs to the site, facilitating problem solving. *Convolutional Siamese neural networks, matching networks, relation networks and prototypical networks* are characteristic metrics-based meta-learning techniques.
c. Optimization-based: These are techniques based on optimizing the parameters of the model for quick learning. *LSTM Meta-Learners, temporal discreteness and the reptile plus Model-Agnostic Meta-Learning (MAML)* algorithms are typical cases of optimization-based meta-Learning.

Table 1. Meta-learning approaches.

	Model-Based	Metric-Based	Optimization-Based
Key idea	RNN; memory	Metric learning	Gradient descent
How is $P_\theta(y\|x)$ modeled?	$F_\theta(x, S)$	$\sum_{(x_i, y_i) \in S} k_\theta(x, x_i) y_i$	$P_{g_\varphi(\theta, S^L)}(y\|x)$

k_θ is a kernel function which calculates the similarity between x_i and x.

The *Recurrent Neural Networks* (RNNs) which use only internal memory, and also the *Long-Short-Term Memory* approaches (LSTM), are not considered meta-learning techniques [11]. Meta-learning can be achieved through a variety of learning examples. In this case, the *supervised gradient-based* learning can be considered as the most effective method [11]. More specifically, the *gradient-based, end-to-end differentiable meta-learning*, provides a wide framework for the application of effective *meta-learning* techniques.

This research proposes an *optimization-based, gradient-based, end-to-end* differentiable meta-learning architecture, based on an innovative evolution of the MAML algorithm [10]. MAML is one of the most successful and at the same time simple optimization algorithms which belongs to the meta-learning approach. One of its great advantages is that it is compatible with any model which learns through the Gradient Descent (GRD) method. It is comprised of the Base-Learner (BL) and the Meta-Learner (ML) models, with the second used to train the first. The weights of the BL are updated following the GRD method in learning tasks of the k-shot problem, whereas the ML applies the GRD approach on the weights of the BL, before the GRD [10].

In Figure 1 you can see a depiction of the MAML algorithm.

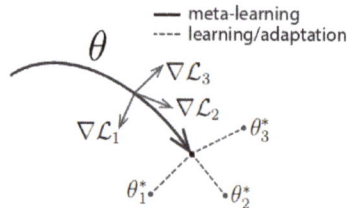

Figure 1. Model-Agnostic Meta-Learning algorithm.

It should be clarified that θ denotes the weights of the meta-learner. Gradient L_i comprises of the losses for task i in a meta-batch. The θ_i^* are the optimal weights for each task. It is essentially an optimization procedure on a set of parameters, such that when a slope step is obtained with respect to a particular task i, the respective parameters θ_i^* are approaching their optimal values. Therefore, the goal of this approach is to learn an intrinsic feature which is widely applicable to all tasks of a distribution $p(T)$ and not to a single one. This is achieved by minimizing the total loss across tasks sampled from the distribution $p(T)$.

In particular, we have a base-model represented by a parametric function f_θ with parameters θ_i and a task $T_i \sim p(T)$. After applying the Gradient Descent, a new feature vector is obtained denoted as θ_i':

$$\theta_i' = \theta - \alpha \nabla_\theta L_{T_i}(f_\theta) \qquad (8)$$

We will consider that we execute only one GD step. The meta-learner optimizes the new parameters using the initial ones, based on the performance of the $f_{\theta'}$ model, in tasks which use sampling from the $P(T)$. Equation (9) is the meta-objective [10]:

$$\min_\theta \sum_{T_i \sim p(T)} L_{T_i}(f_{\theta_i'}) = \min_\theta \sum_{T_i \sim p(T)} L_{T_i}(f_\theta - \alpha \nabla_\theta L_{T_i}(f_\theta)) \qquad (9)$$

The meta-optimization is performed again with the *Stochastic Gradient Descent* (SGD) and it updates the parameters θ as follows:

$$\theta \leftarrow \theta - \beta \nabla_\theta \sum_{T_i \sim p(T)} L_{T_i}(f_{\theta_i'}) \qquad (10)$$

It should be noted that we do not actually define an additional set of variables θ_i' whose values are calculated by considering one (or more) Gradient Descents from θ relative to *task i*. This step is known as the *Inner Loop Learning* (INLL), which is the reverse process of the *Outer Loop Learning* (OLL), and it optimizes Equation (10). If, for example, we apply INLL to fine-tune θ for process i, then according to Equation (10) we are optimizing a target with the expectation which the model applies to each task, following corresponding fine-tuning procedures.

The following Algorithm 1 is an analytical presentation of the MAML algorithm [10].

Algorithm 1. MAML.

Require: $p(T)$: distribution over tasks
Require: α, β: step size hyperparameters
1: randomly initialize θ
2: **while** not done **do**
3: Sample batch of tasks $T_i \sim p(T)$
4: **for all** T_i **do**
5: Evalluate $\nabla_\theta L_{T_i}(f_\theta)$ with respect to K examples
6: Compute adapted parameters with gradient descent: $\theta'_i = \theta - \alpha \nabla_\theta L_{T_i}(f_\theta)$
7: **end for**
8: Update $\theta \leftarrow \theta - \beta \nabla_\theta \sum_{T_i \sim p(T)} L_{T_i}(f_{\theta'_i})$
9: **end while**

Figure 2 is a graphical illustration of the operation of the MAML algorithm.

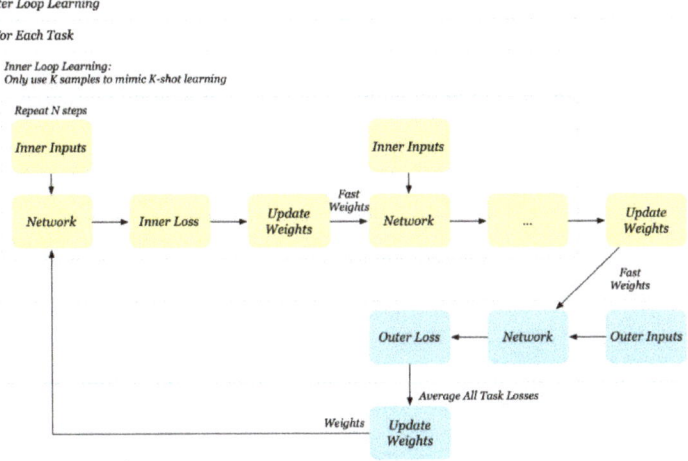

Figure 2. Graphical presentation of the MAML.

It should be clarified that the intermediate parameters θ'_i are considered fast weights. The INLL considers all of the N gradient steps for the final estimation of the fast weights, based on the fact that the outer learning loop calculates the outer task loss $L_{T_i}(f_{\theta'_i})$. However, though the inner learning loop makes N iterations, the MAML algorithm employs only the final weights to perform the OLL. However, this is a fairly significant problem, as it can create instability in learning when N is large.

The field of few-shot or Zero-shot Learning, has recently seen substantial advancements. Most of these advancements came from casting few-shot learning as a meta-learning problem. MAML is currently one of the best approaches for few-shot learning via meta-learning. It is a simple, general, and effective optimization algorithm that does not place any constraints on the model architecture or loss functions. As a result, it can be combined with arbitrary networks and different types of loss functions, which makes it applicable to a variety of different learning processes. However, it has a variety of issues, such as being very sensitive to neural network architectures, often leading to instability during training. It requires arduous hyperparameter searches to stabilize training and achieve high generalization, and it is very computationally expensive at both training and inference times.

3. Related Work

Although the MAML algorithm and its variants do not use parameters other than those of the base-learner, network training is quite slow and computationally expensive as it contains second-degree derivatives. In particular, the meta-update of the MAML algorithm includes gradient nested in gradient, or second-degree derivatives, which significantly increases the computational cost. In order to solve the above problem, several approximation techniques have been proposed to accelerate the algorithm.

Finn et al. [12] developed the MAML by ignoring the second derivatives, calculating the slope in the meta-update, which they called FOMAML (First Order MAML).

More specifically, MAML optimizes the:

$$\min_{\theta} \mathbb{E}_{T \sim p(T)} \left[L_T \left(\mathbb{U}_T^k(\theta) \right) \right], \tag{11}$$

where $\mathbb{U}_T^k$ is the process by which k samples are taken from task T and θ is updated. This procedure employs the support set and the query set, so the optimization can be rewritten as follows:

$$\min_{\theta} \mathbb{E}_{T \sim p(T)} \left[L_{T,Q}(\mathbb{U}_{T,S}(\theta)) \right] \tag{12}$$

Finally, MAML uses the slope method to calculate the following:

$$gMAML = L_{T,Q}(\mathbb{U}_{T,S}(\theta)) = \mathbb{U}'_{T,S}(\theta) L'_{T,Q}(\widetilde{\theta}), \tag{13}$$

where $\widetilde{\theta} = \mathbb{U}_{T,S}(\theta)$ and $\mathbb{U}'_{T,S}$ is the Jacobian renewal matrix of $\mathbb{U}_{T,S}$ where the FOMAML considers $\mathbb{U}'_{T,S}$ as unitary, so it calculates the following:

$$gFOMAML = L'_{T,Q}(\widetilde{\theta}) \tag{14}$$

The resulting method still calculates the meta-gradient for the parameter values after updating θ', which is an effective post-learning method from a theoretical point of view. However, experiments have shown that the yield of this method is almost the same as the one obtained by a second derivative. Most of the improvement in MAML comes from the gradients of the objective at the post-update parameter values, rather than the second-order updates from differentiating through the gradient update.

A different implementation employing first degree derivatives was studied and analyzed by Nichol et al. [13]. They introduced the reptile algorithm, which is a variation of the MAML, using only the first derivative. The basic difference from FOMAML is that the last step treats $\widetilde{\theta} - \theta$ as a slope and feeds it into an adaptive algorithm such as ADAM. Algorithm 2 presents reptile.

Algorithm 2. Reptile algorithm.

Initialize θ, the vector of initial parameters
1: **for** *iteration* $= 1, 2, \ldots,$ **do**
2: Sample task T, corresponding to Loss L_T on weighs vector $\widetilde{\theta}$
3: Compute $\widetilde{\theta} = \mathbb{U}_T^k(\theta)$, denoting k steps of gradient descent or Adam algorithm
4: Update $\theta \leftarrow \theta + \varepsilon(\widetilde{\theta} - \theta)$
5: **end for**

MAML also suffers from training instability, which can currently only be alleviated by arduous architecture and hyperparameter searches.

Antoniou et al. proposed an improved variant of the algorithm, called MAML++, which effectively addresses MAML problems, by providing a much improved training stability and removing the dependency of training stability on the model's architecture. Specifically, Antoniou et al. [14] found that simply replacing max-pooling layers with stridden convolutional layers makes network training unstable. It is clearly shown in Figure 3 that in two of the three cases, the original MAML appears to be unstable and irregular, while all 3 MAML++ models appear to converge consistently very quickly, with much higher generalization accuracy without any stability problems.

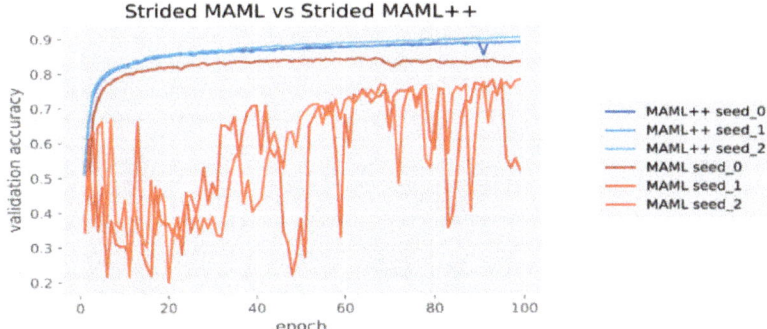

Figure 3. Stabilizing MAML.

It was estimated that the instability was caused by a gradient degradation (gradient explosion or vanishing gradient) which was due to the depth of the network. Let us consider a typical four-layer Convolutional Neural Network (CNN) followed by a single linear layer. If we repeat the Inner Loop Learning N times, then the inference graph comprises 5N layers in total, without any skip connections.

Since the original MAML only uses the final weights for the Outer Loop Learning, the backpropagation algorithm has to go through all layers, which causes gradient degradation. To solve the above problem, the Multi-Step Loss (MSL) optimization approach was adopted. It eliminates the problem by calculating the external loss after each internal step, based on the outer loop update, as in Equation (15) below:

$$\theta = \theta - \beta \nabla_\theta \sum_{i=1}^{B} \sum_{j=1}^{N} w_j L_{T_i}\left(f_{\theta_j^i}\right), \tag{15}$$

where β is a Learning Rate; $L_{T_i}(f_{\theta_j^i})$ denotes the outer loss of task i when using the base-network weights after the j-inner-step update; and w_j denotes the importance weight of the outer loss at step j.

The following, Figure 4, is a graphical display of the MAML++ algorithm, where the outer loss is calculated after each internal step and the weighted average is obtained at the end of the process.

In practice, all losses are initialized with equal contributions to the overall loss, but as repetitions increase, contributions from previous steps are reduced, while the ones of subsequent steps keep increasing gradually. This is to ensure that as training progresses, the final step loss receives more attention from the optimizer, thereby ensuring that the lowest possible loss is achieved. If the annealing is not used, the final loss might be higher compared to the one obtained by the original formulation. Additionally, due to the fact that the original MAML overcomes the second-order derivative cost by completely ignoring it, the final general performance of the network is reduced. The MAML++ solves this problem, by using the *derivative order annealing* method. Specifically, it employs the first order grade for the first 50 epochs of training and then it moves to the second order grade for the rest of the training process. An interesting observation is that this derivative-order annealing does not create incidents of exploding or diminishing gradients, and so the training is much more stable.

Another drawback of MAML is the fact that it does non-use *batch normalization statistic accumulation*. Instead, only the statistics of the current batch are used. As a result, smoothing is less effective, as the trained parameters must include a variety of different means and standard deviations from different tasks. A naive application would accumulate current batch statistics at all stages of the Inner Loop Learning update, which could cause optimization problems, and it could slow or stop optimization altogether. The problem stems from the erroneous assumption that the original model and all its updated iterations have similar attribute distributions. Thus, the current statistics could be shared across all updates to the internal loop of the network. Obviously, this assumption is not correct. A better alternative solution, which is employed by MAML++, is the storage of *per-step batch normalization statistics* and the reading of *per-step batch normalization weights and* biases for each repetition of the inner loop. One issue that affects the speed of generalization and convergence is the use of a *shared Learning Rate* for all parameters and all steps of learning process update.

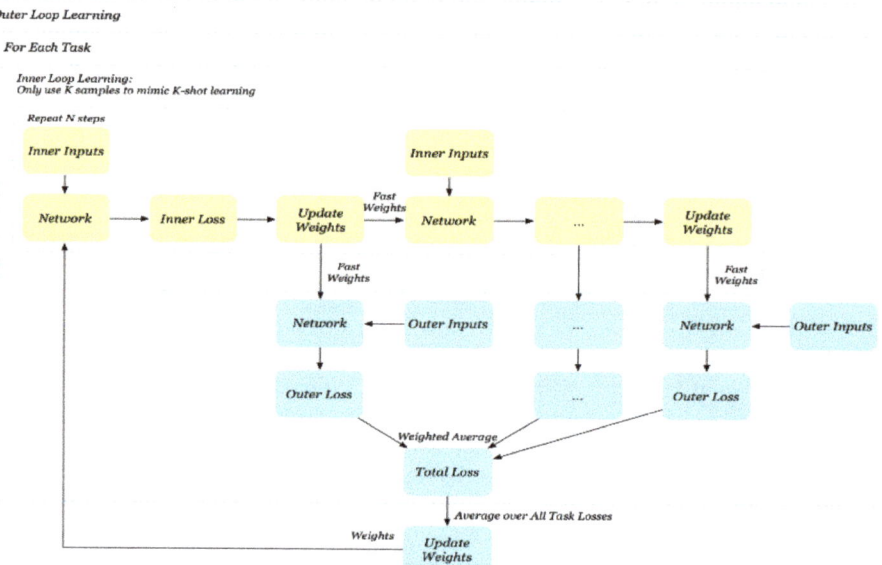

Figure 4. MAML++ visualization.

The consistent Learning Rate requires multiple hyperparameter searches, in order to find the right rate for a particular set of data. This is computationally expensive and time consuming, depending on how the search is shared. Use the shared Learning Rate for all parameters and all the steps of updating the learning process. In addition, while gradient is an effective data fitting tactic, a constant Learning Rate can easily lead to overfitting under the *few-shot regime*. An approach to avoid potential overfitting is the identification of all learning factors in a way that maximizes the power of generalization rather than the over-fitting of the data.

Li et al. [15] proposed a Learning Rate for each parameter in the core network where the internal loop was updated, as in the following equation (Equation (16)):

$$\theta' = \theta - \alpha \circ \nabla_\theta L_{T_i}(f_\theta), \tag{16}$$

where α is a vector of learnable parameters with the same size as $L_{T_i}(f_\theta)$ and denotes the element-wise product operation. We do not put the constraint of positivity on the Learning Rate (LER) denoted as "α." Therefore, we should not expect the inner-update direction to follow the gradient direction.

A clearly improved approach to the above process is suggested by MAML++ which employs *per-layer per-step* Learning Rates. For example, if it is assumed that the *core network* comprises L layers and the *Inner Loop Learning* consists of N stages of updating, then there are LN additional learnable parameters for the *Inner Loop Learning Rate*.

MAML uses the *ADAM* algorithm with a constant LER to optimize the meta-objective. This means which more time is required to properly adjust the Learning Rate, which is a critical parameter of the generalization performance. On the other hand, MAML++ employs the cosine annealing scheduling on the *meta-optimizer*, which is defined based on the following Equation (17) [16].

$$\beta = \beta_{min} + \frac{1}{2}(\beta_{max} - \beta_{min})(1 + \cos(\frac{T}{T_{max}}\pi)), \qquad (17)$$

where β_{min} denotes the minimum Learning Rate, β_{max} denotes the initial Learning Rate, T is the number of current iterations and T_{max} is the maximum number of iterations. When $T = 0$, the LER $\beta = \beta_{max}$. On the other hand, if $T = T_{max}$, then $\beta = \beta_{min}$. In practice, we might want T to be T_{max}.

In summary, this particular MAML++ standardization enables its use in complex Deep Learning architectures, making it easier to learn more complex functions, such as loss functions, optimizers or even gradient computation functions. Moreover, the use of first-class derivatives offers a powerful pre-training method aiming to detect the parameters which are less likely to cause exploding or diminishing gradients. Finally, the *learning per-layer per-step* LER technique avoids potential overfitting, while it significantly reduces the computational cost and time required to build a consistent Learning Rate throughout the process.

4. Design Principles and Novelties of the Introduced MAME-ZsL Algorithm

As it has already been mentioned, the proposed MAME-ZsL algorithm employs MAML++ for the development of a robust *Hyperspectral Image Analysis* and *Classification* (HIAC) model, based on ZsL. The basic novelty introduced by the improved MAME-ZsL model, is related to a neural network with Convolutional (CON) filters, comprising very small receptive fields of size 3 × 3.

The Convolutional stride and the spatial padding were set to 1 pixel. Max-pooling was performed over 3 × 3 pixels windows with stride equal to three. All of the CON layers were developed using the Rectified Linear Unit (ReLU) nonlinear Activation Function (ACF), except for the last layer where the Softmax ACF [3] was applied, as it performs better on multi-classification problems like the one under consideration (18).

$$\sigma_j(z) = \frac{e^{z_j}}{\sum_{k=1}^{K} e^{z_k}}, \, j = 1, \ldots, K \qquad (18)$$

The Sigmoid approach offers better results in binary classification tasks. It is a fact that in the Softmax, the sum of probabilities is equal to 1, which is not the case for the Sigmoid. Moreover, in Softmax the highest value has a higher probability than the others, while in the Sigmoid the highest value is expected to have a high probability but not the highest one.

The fully Convolutional Neural Network (CNN) was trained based on the novel *AdaBound* algorithm [17] which employs dynamic bounds on the Learning Rate and it achieves a smooth transition to stochastic gradient descent. Algorithm 3 makes a detailed presentation of the AdaBound [17]:

Algorithm 3. The AdaBound algorithm.

Input: $x_1 \in F$, initial step size α, $\{\beta_{1t}\}_{t=1}^{T}$, β_2 lower bound function η_l, upper bound function η_u
1: Set $m_0 = 0$, $u_0 = 0$
2: **for** $t = 1$ to T **do**
3: $\quad g_t = \nabla f_t(x_t)$
4: $\quad m_t = \beta_{1t} m_{t-1} + (1 - \beta_{1t}) g_t$
5: $\quad u_t = \beta_2 u_{t-1} + (1 - \beta_2) g_t^2$ and $V_t = diag(u_t)$
6: $\quad \hat{\eta}_t = Clip\left(\frac{\alpha}{\sqrt{V_t}}, \eta_l(t), \eta_u(t)\right)$ and $\eta_t = \frac{\hat{\eta}_t}{\sqrt{t}}$
7: $\quad x_{t+1} = \prod_{f, diag(\eta_t^{-1})} (x_t - \eta_t \cdot m_t)$
8: **end for**

Compared to other methods, AdaBound has two major advantages. It is uncertain whether there exists a fixed turning point to distinguish the simple ADAM algorithm and the SGD. The advantage of the AdaBound is the fact that it addresses this problem with a continuous transforming procedure, rather than with a "hard" switch. The AdaBound introduces an extra hyperparameter to perform the switching time, which is not very easy to fine-tune. Moreover, it has a higher convergence speed than the stochastic gradient descent ones. Finally, it overcomes the poor generalization ability of the adaptive models, as it uses dynamic limits on the LER, aiming towards higher classification accuracy.

The selection of the appropriate hyperparameters to be employed in the proposed method, was based on the restrictions' settings and configurations, which should be based on the consideration of the different decision boundaries of the classification problem. For example, the obvious choice of classifiers with the smallest error in training data is considered as improper for generating a classification model. The performance based on a training dataset, even when cross-validation is used, may be misleading when first seen data vectors are used. In order for the proposed process to be effective, individual hyperparameters were chosen. They not only display a certain level of diversity, but they also use different operating functions, thus allowing different decision boundaries to be created and combined in such a way that can reduce the overall error.

In general, the selection of features was based on a heuristic method which considers the way the proposed method faces each situation. For instance:

- Are any parametric approaches employed?
- What is the effect of the outliers? (The use of a subgroup of training sets with bagging can provide significant help towards the reduction of the effect of outliers or extreme values)
- How is the noise handled? For instance: if it is repeatedly non-linear, it can detect linear or non-linear dispersed data; it tends to perform very well with a lot of data vectors. The final decision is made based on the performance encountered by the statistical trial and error method.

5. Application of the MAME-ZsL in Hyperspectral Image Analysis

Advances in artificial intelligence, combined with the extended availability of high quality data and advances in both hardware and software, have led to serious developments in the efficient processing of data related to the GeoAI field (*Artificial Intelligence and Geography/Geographic Information Systems*). Hyperspectral Image Analysis for efficient and accurate object detection using Deep Learning is one of the timely topics of GeoAI. The most recent research examples include detection of soil characteristics [18], detailed ways of capturing densely populated areas [19], extracting information from scanned historical maps [20], semantic point sorting [21], innovative spatial interpolation methods [22] and traffic forecasting [23].

Similarly, modern applications of artificial vision and imaging (IMG) systems significantly extend the distinctive ability of optical systems, both in terms of spectral sensitivity and resolution. Thus, it is possible to identify and differentiate spectral and spatial regions, which although having the same color appearance, are characterized by different physico-chemical and/or structural properties. This differentiation is based on the emerging spatial diversification, which is detected by observing in narrow spectral bands, within or outside the visible spectrum.

Recent technological developments have made it possible to combine IMG (spatial variation in RGB resolution) and spectroscopy (spectral analysis in spatially emitted radiation) in a new field called "Spectral Imaging" (SIM). In the SIM process, the intensity of light is recorded simultaneously as a function of both wavelength and position. The dataset corresponding to the observed surface contains a complete image, different for each wavelength. In the field of spectroscopy, a fully resolved spectrum can be recorded for each pixel of the spatial resolution of the observation field. The multitude of spectral regions, which the IMG system can manage, determines the difference between multispectral (tens of regions) and Hyperspectral (hundreds of regions) Imaging. The key element of a typical spectral IMG system is the monochromatic image sensor (monochrome camera), which can be used to select the desired observation wavelength.

It can be easily perceived that the success of a sophisticated Hyperspectral Analysis System (HAS) is a major challenge for DL technologies, which are using a series of algorithms attempting to model data characterized by high-level of abstraction. HAS use a multi-level processing architecture, which is based on sequential linear and non-linear transformations. Despite their undoubtedly well-established and effective approaches and their advantages, these architectures depend on the performance of training with huge datasets which include multiple representations of images of the same class. Considering the multitude of classes which may be included in a Hyperspectral image, we realize that this process is so incredibly time consuming and costly, that it can sometimes be impossible to run [1].

The ZsL method was adopted based on a heuristic [24], hierarchical parameter search methodology [25]. It is part of a family of learning techniques which exploit data representations to interpret and derive the optimal result. This methodology uses distributed representation, the basic premise of which is that the observed data result from the interactions of factors which are organized in layers. A fundamental principle is that these layers correspond to levels of abstraction or composition based on their quantity and size.

Fine-Grained Recognition (FIG_RC) is the task of distinguishing between visually very similar objects, such as identifying the species of a bird, the breed of a dog or the model of an aircraft. On the other hand, FIG_RC [26] which aims to identify the type of an object among a large number of subcategories, is an emerging application with the increasing resolution which exposes new details in image data. Traditional fully supervised algorithms fail to handle this problem where there is low between-class variance and high within-class variance for the classes of interest with small sample sizes. The experiments show that the proposed fine-grained object recognition model achieves only 14.3% recognition accuracy for the classes with no training examples. This is slightly better than a random guess accuracy of 6.3%. Another method [27] automatically creates a training dataset from a single degraded image and trains a denoising network without any clear images. However, the performance of the proposed method shows the same performance as the optimization-based method at high noise levels.

Hu et al., 2015, proposed a time-consuming and resource depended model [28] which learns to perform zero-shot classification, using a meta-learner which is trained to produce corrections to the output of a previously trained learner. The model consists of a Task Module (TM) which supplies an initial prediction, and a Correction Module (CM) updating the initial prediction. The TM is the learner and the CM is the meta-learner. The correction module is trained in an episodic approach, whereby many different task modules are trained on various subsets of the total training data, with the rest being used as unseen data for the CM. The correction module takes as input a representation of the TM's training data to perform the predicted correction. The correction module is trained to update the task module's prediction to be closer to the target value.

In addition [29] proposes the use of the visual space as the embedding one. In this space, the subsequent nearest neighbor search suffers much less from the harness problem and it becomes more effective. This model design also provides a natural mechanism for multiple semantic modalities (e.g., attributes and sentence descriptions) to be fused and optimized jointly in an end-to-end manner. Only the statistics of the current environment are used and the trained process must include a variety of different statistics from different tasks and environments.

Additionally, [30] propose a very promising approach with high-grade accuracy, but the model is characterized by high bias. In the case of image classification, various spatial information can be extracted and used, such as edges, shapes and associated color areas. As they are organized into multiple levels, they are hierarchically separated into levels of abstraction, creating the conditions for selecting the most appropriate features for the training process. Utilizing the above processes, ZsL inspires and simulates the functions of human visual perception, where multiple functional levels and intermediate representations are developed, from capturing an image to the retina to responding in stimuli. This function is based on the conversion of the input representation to a higher level one, as it is performed by each intermediate unit. High-level features are more general and unchanged, while low-level ones help to categorize inputs. Their effectiveness is interpreted on the basis of the *"universal approximation theorem,"* which deals with the ability of a neural structure to approach continuous functions and the probabilistic inference which considers the activation of nonlinearity as a function of cumulative distribution. This is related to the concepts of optimization and generalization respectively [25].

Given that in deep neural networks, each hidden level trains a distinct set of features, coming from the output of the previous level, further operation of this network enables the analysis of the most complex features, as they are reconstructed and decomposed from layer to layer. This hierarchy, as well as the degradation of information, while increasing the complexity of the system, also enables the handling of high-dimensional data, which pass through non-linear functions. It is thus possible to discover unstructured data and to reveal a latent structure in unmarked data. This is done in order to handle more general problematic structures, even discerning the minimal similarities or anomalies they entail.

Specifically, since the aim was the design of a system with zero samples from the target class, the proposed methodology used the intermediate representations extracted from the rest of the image samples. This was done in order to find the appropriate representations to be used in order to classify the unknown image samples.

To increase the efficiency of the method, bootstrap sampling was used, in order to train different subsets of the data set in the most appropriate way. Bootstrap sampling is the process of using increasingly larger random samples until the accuracy of the neural network is improved. Each sample is used to compile a separate model and the results of each model are aggregated with "voting"; that is, for each input vector, each classifier predicts the output variable, and finally, the value with the most "votes" is selected as the response variable for which particular vector. This methodology, which belongs to the ensemble methods, is called bagging and has many advantages, such as reducing co-variance and avoiding overfitting, as you can see in the below Figure 5 [31].

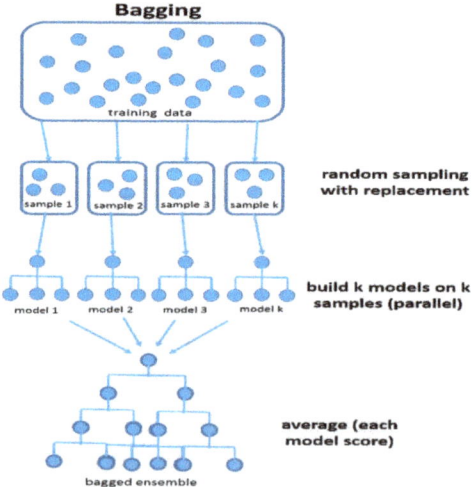

Figure 5. Bagging (bootstrap sampling) (https://www.kdnuggets.com/).

The ensemble approach was selected to be employed in this research, due to the particularly high complexity of the examined ZsL, and due to the fact that the prediction results were highly volatile. This can be attributed to the sensitivity of the correlational models to the data, and to the complex relationship which describes them. The ensemble function of the proposed system offers a more stable model with better prediction results. This is due to the fact that the overall behavior of a multiple model is less noisy than a corresponding single one. This reduces the overall risk of a particularly bad choice.

It is important to note that in Deep Learning, the training process is based on analyzing large amounts of data. The research and development of neural networks is flourishing thanks to recent advancements in computational power, the discovery of new algorithms and the increase in labeled data.

Neural networks typically take longer to run, as an increase in the number of features or columns in the dataset also increases the number of hidden layers. Specifically, we should say that a single affine layer of a neural network without any non-linearities/activations is practically the same as a linear model. Here we are referring to deep neural networks that have multiple layers and activation functions (non-linearities as Relu, Elu, tanh, Sigmoid) Additionally, all of the nonlinearities and multiple layers introduce a nonconvex and usually rather complex error space, which means that we have many local minimums that the training of the deep neural network can converge to. This means that a lot of hyperparameters have to be tuned in order to get to a place in the error space where the error is small enough so that the model will be useful. A lot of hyper parameters which could start from 10 and reach up to 40 or 50 are dealt with via Bayesian optimization, using Gaussian processes to optimize them, which still does not guarantee good performance. Their training is very slow, and adding the tuning of the hyperparameters into that makes it even slower, whereas the linear model would be much faster to be trained. This introduces a serious cost–benefit tradeoff. A trained linear model has weights which are interpretable and gives useful information to the data scientist as to how various features should have roles for the task at hand.

Modern frameworks like TensorFlow or Theano perform execution of neural networks on GPU. They take advantage of parallel programming capabilities for large array multiplications, which are typical of backpropagation algorithms.

The proposed Deep Learning model is a quite resource-demanding technology. It requires powerful, high-performance graphics processing units and large amounts of storage to train the models. Furthermore, this technology needs more time to train in comparison with traditional machine learning. Another important disadvantage of any Deep Learning model is that it is incapable of providing arguments about why it has reached a certain conclusion. Unlike in the case of traditional machine learning, you cannot follow an algorithm to find out why your system has decided which it is a tree on a picture, not a tile. To correct errors in Deep Learning, you have to revise the whole algorithm.

6. Description of the Datasets

The datasets used in this research include images taken from a Reflective Optics System Imaging Spectrometer (ROSIS). More specifically, the Pavia University and Pavia Center datasets were considered [32]. Both datasets came from the ROSIS sensor during a flight campaign over Pavia in southern Italy. The number of spectral bands is 102 for the Pavia Center and it is 103 for Pavia University. The selected Pavia Center and Pavia University images have an analysis of 1096 × 1096 pixels and 610 × 610 pixels respectively. Ultrasound imaging consists of 115 spectral channels ranging from 430 to 860 nm, of which only 102 were used in this research, as 13 were removed due to noise. Rejected specimens which in both cases contain no information (including black bars) can be seen in the following figure (Figure 6) below.

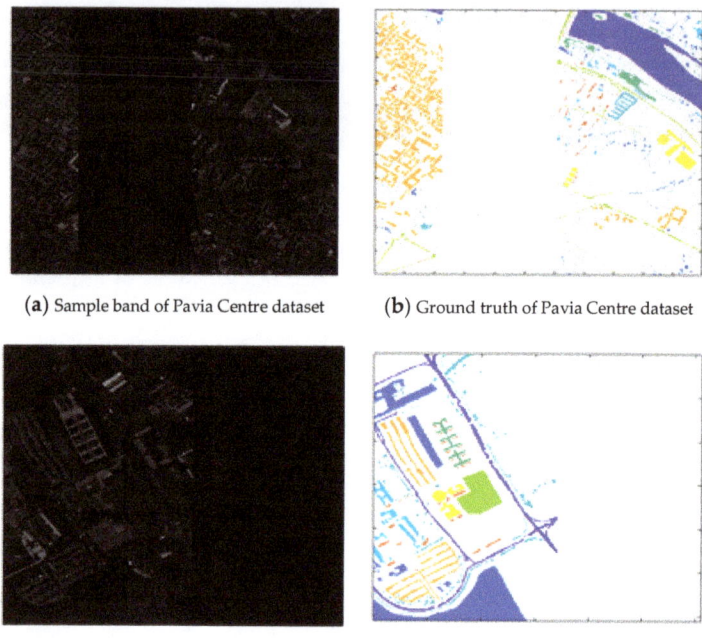

(**a**) Sample band of Pavia Centre dataset (**b**) Ground truth of Pavia Centre dataset

(**c**) Sample band of Pavia University dataset (**d**) Ground truth of Pavia University dataset

Figure 6. Noisy bands in Pavia Centre and University datasets.

The available samples were scaled down, so that every image has an analysis of 610 × 610 pixels and geometric analysis of 1.3 m. In both datasets the basic points of the image belong to nine categories which are mainly related to land cover objects. The Pavia University dataset includes nine classes, and in total, 46,697 cases. The Pavia Center dataset comprises nine classes with 7456 cases, whereas the first seven classes are common in both datasets (*Asphalt, Meadows, Trees, Bare Soil, Self-Blocking Bricks, Bitumen, Shadows*) [33].

The Pavia University dataset was divided into training, validation and testing sets, as is presented in the following table (Table 2) [32].

Table 2. Pavia University dataset.

Class No	Class Name	All Instances	Sets		
			Training	Validation	Test
1	Asphalt	7179	√	-	-
2	Meadows	19,189	√	-	-
3	Trees	3588	√	-	-
4	Bare Soil	5561	√	-	-
5	Self-Blocking Bricks	4196	√	-	-
6	Bitumen	1705	√	-	-
7	Shadows	1178	√	-	-
8	Metal Sheets	1610	-	√	-
9	Gravel	2491	-	-	√
	Total	46,697	42,596	1610	2491

The Pavia Center dataset was divided into training, validation and testing sets, as is presented analytically in the following table (Table 3) [32].

Table 3. Pavia Center dataset.

Class No	Class Name	All Instances	Sets		
			Training	Validation	Test
1	Asphalt	816	√	-	-
2	Meadows	824	√	-	-
3	Trees	820	√	-	-
4	Bare Soil	820	√	-	-
5	Self-Blocking Bricks	808	√	-	-
6	Bitumen	808	√	-	-
7	Shadows	476	√	-	-
8	Water	824	-	√	-
9	Tiles	1260	-	-	√
	Total	7456	5372	824	1260

Metrics Used for the Assessment of the Modeling Effort

The following metrics were used for the assessment of the modeling effort [33,34]:

(a) Overall Accuracy (OA): This index represents the number of samples correctly classified, divided by the number of testing samples.

(b) Kappa Statistic: This is a statistical measure which provides information on the level of agreement between the truth map and the final classification map. It is the percentage of agreement corrected by the level of agreement, which could be expected to occur by chance. In general, it is considered to be a more robust index than a simple percent agreement calculation, since k takes into account the agreement occurring by chance. It is a popular measure for benchmarking classification accuracy under class imbalance. It is used in static classification scenarios and for streaming data classification. Cohen's kappa measures the agreement between two raters, where each classifies N items into C mutually exclusive categories. The definition of κ is [35,36]:

$$\kappa = \frac{p_o - p_e}{1 - p_e} = 1 - \frac{1 - p_o}{1 - p_e}, \quad (19)$$

where p_o is the relative observed agreement among raters (identical to accuracy), and p_e is the hypothetical probability of chance agreement. The observed data are used to calculate the probabilities of each observer, to randomly see each category. If the raters are in complete agreement, then $\kappa = 1$. If there is no agreement among the raters other than what would be expected by chance (as given by p_e), then $\kappa \approx 0$.

The Kappa Reliability (KR) can be considered as the outcome from the data editing, allowing the conservancy of more relevant data for the upcoming forecast. A detailed analysis of the KR is presented in the following Table 4.

Table 4. Kappa Reliability.

Kappa	Reliability
0.00	no reliability
0.1–0.2	minimum
0.21–0.40	little
0.41–0.60	moderate
0.61–0.80	important
≥0.81	maximum

(c) McNemar test: The McNemar statistical test was employed to evaluate the importance of classification accuracy derived from different approaches [31]:

$$z_{12} = \frac{f_{12} - f_{21}}{\sqrt{f_{12} + f_{21}}},\qquad(20)$$

where f_{ij} is the number of correctly classified samples in classification i, and incorrect one are in classification j. McNemar's test is based on the standardized normal test statistic, and therefore the null hypothesis, which is "no significant difference," rejected at the widely used $p = 0.05$ ($|z| > 1.96$) level of significance.

7. Results and Comparisons

The training of the models was performed using a Learning Rate of 0.001. Loss function is the cross-entropy error between the predicted and true class. The cross-entropy error was used as the Loss function between the predicted and the true class. For the other parameters of the model, the recommended default settings were set as in [37].

A comparison with the following most widely used supervised Deep Learning models was performed to validate the effectiveness of the proposed architecture.

(a) 1-D CNN: The architecture of the 1-D CNN was designed as in [38], and it comprises the input layer, the Convolutional Layer (COL), the max-pooling layer, the fully connected layer and the output one. The number of the Convolutional Filters (COFs) was equal to 20, the length of each filer was 11 and the pooling size had the value of 3. Finally, 100 hidden units were contained in the fully connected layer.

(b) 2-D CNN: The 2-D CNN architecture was designed as in [39]. It includes three COLs which were supplied with 4×4, 5×5 and 4×4 CON filters respectively. The COL except of the final one were followed by max-pooling layers. Moreover, the numbers of COFs for the COLs were to 32, 64 and 128, respectively.

(c) Simple Convolutional/deconvolutional network with the simple Convolutional blocks and the unpooling function, as it is described in [40,41].

(d) Residual Convolutional/deconvolutional network: Its architecture used residual blocks and a more accurate unpooling function, as it is shown in [42].

The following Tables 5 and 6 show the classification maps which have emerged for the cases of the Pavia University and Pavia Center datasets. Moreover, they present a comparative analysis of the accuracy of the proposed MAME-ZsL algorithm, with the performance of the following classifiers; namely: 1-D CNN, 2-D CNN, Simple Convolutional/Deconvolutional Network (SC/DN), Residual Convolutional/Deconvolutional Network (RC/DN).

A single dataset including all records of the Pavia University and Pavia Center datasets was developed, which was employed in an effort to fully investigate the predictive capacity of the proposed system. It was named *General Pavia Dataset (GPD)*, and it was divided into training (Classes 1–7), validation (Classes 8 and 9) and testing subsets (Classes 10 and 11). It is presented in the following Table 7.

Table 5. Testing classification accuracy and performance metrics.

Class Name	Pavia University Dataset													
	1-D CNN			2-D CNN			SC/DN			RC/DN			MAME-ZsL		
	OA	κ	McN	OA	κ	McN	OA	κ	McN	OA	κ	McN	OA	κ	McN
Metal Sheets	99.41%	0.8985	33.801	100%	1	32.752	97.55%	0.8356	30.894	97.77%	0.8023	29.773	78.56%	0.7292	30.856
Gravel	67.03%	0.7693		63.13%	0.7576		60.31%	0.7411		61.46%	0.7857		54.17%	0.7084	

Table 6. Testing classification accuracy and performance metrics.

Class Name	Pavia Center Dataset													
	1-D CNN			2-D CNN			SC/DN			RC/DN			MAME-ZsL		
	OA	κ	McN	OA	κ	McN	OA	κ	McN	OA	κ	McN	OA	κ	McN
Water	77.83%	0.8014	32.587	79.97%	0.8208	32.194	80.06%	0.8114	31.643	82.77%	0.8823	30.588	62.08%	0.7539	31.002
Tiles	81.15%	0.8296		76.72%	0.7994		80.67%	0.7978		78.34%	0.8095		65.37%	0.7111	

Table 7. The general Pavia dataset.

Class No	Class Name	All Instances	Sets		
			Training	Validation	Test
1	Asphalt	7995	✓	-	-
2	Meadows	20,013	✓	-	-
3	Trees	4408	✓	-	-
4	Bare Soil	6381	✓	-	-
5	Self-Blocking Bricks	5004	✓	-	-
6	Bitumen	2513	✓	-	-
7	Shadows	1654	✓	-	-
8	Metal Sheets	1610	-	✓	-
9	Water	824	-	✓	-
10	Gravel	2491	-	-	✓
11	Tiles	1260	-	-	✓
	Total	47,968	42,596	2434	3751

As it can be seen from Table 8, the increase of samples has improved the results in the case of the trained algorithms and in the case of the ZsL technique. It is easy to conclude which the proposed MAME-ZsL is a highly valued Deep Learning system which has achieved remarkable results in all evaluations over their respective competing approaches.

The experimental comparison does not include other examples of Zero-shot Learning. This fact does not detract in any case from the value of the proposed method, taking into account that the proposed processing approach builds a predictive model which is comparable to supervised learning systems. The appropriate hyperparameters in the proposed method are also identified by the high reliability and overall accuracy on the obtained results. The final decision was taken based on the performance encountered by the statistical trial and error method. The performance of the proposed model was evaluated against state-of-the-art fully supervised Deep Learning models. It is worth mentioning that the proposed model was trained with instances only from seven classes while the Deep Learning models were trained with instances from all eleven classes. The presented numerical experiments demonstrate that the proposed model produces remarkable results compared to theoretically superior models, providing convincing arguments regarding the classification efficiency of the proposed approach. Another important observation is that it produces accurate results without recurring problems of undetermined cause, because all of the features in the considered dataset are efficiently evaluated. The values of the obtained kappa index are a proof of high reliability (the reliability can be considered as high when $\kappa \geq 0.70$) [35,36].

The superiority of the introduced, novel model focuses on its robustness, accuracy and generalization ability. The overall behavior of the model is comparable to a corresponding supervised one. Specifically, it reduces the possibility of overfitting, it decreases variance or bias and it can fit unseen patterns without reducing its precision. This is a remarkable innovation which significantly improves the overall reliability of the modeling effort. This is the result of the data processing methodology which allows the retention of the more relevant data for upcoming forecasts.

The following Table 8 presents the results obtained by the analysis of the GPD.

The above, Table 8, also provides information on the results of the McNemar test, to assess the importance of the difference between the classification accuracy of the proposed network and the other approaches examined. The improvement of the overall accuracy values obtained by the novel algorithm (compared to the other existing methods) is statistically significant.

Finally, the use of the ensemble approach in this work is related to the fact that very often in multi-factor problems of high complexity such as the one under consideration, the prediction results show multiple variability [43]. This can be attributed to the sensitivity of the correlation models to the data. The main imperative advantage of the proposed ensemble model is the improvement of the overall predictions and the generalization ability (adaptation in new previously unseen data). The ensemble method definitely decreases the overall risk of a particularly poorer choice.

The employed bagging technique offers better prediction and stability, as the overall behavior of the model becomes less noisy and the overall risk of a particularly bad choice that may be caused from under-sampling is significantly reduced. The above assumption is also supported by the dispersion of the expected error, which is close to the mean error value, something which strongly indicates the reliability of the system and the generalization capability that it presents.

As it can be seen in Tables 5–8, the ensemble method appears to have the same or a slightly lower performance across all datasets, compared to the winning (most accurate) algorithm. The highly overall accuracy shows the rate of positive predictions, whereas k reliability index specifies the rate of positive events which were correctly predicted. In all cases, the proposed model has high average accuracy and very high reliability, which means that the ensemble method is a robust and stable approach which returns substantial results. Tables 7 and 8 show clearly that the ensemble model is very promising.

Table 8. Testing classification accuracy and performance metrics.

| Class Name | General Pavia Dataset (OA Is the Overall Accuracy) ||||||||||||
| | 1-D CNN ||| 2-D CNN ||| SC/DN ||| RC/DN ||| MAME-ZsL |||
	OA	κ	McN	OA	κ	McN	OA	κ	McN	OA	κ	McN	OA	κ	McN
Metal Sheets	99.44%	0.8992		100%	1		99.09%	0.8734		97.95%	0.8137		81.16%	0.8065	
Water	OA	κ		OA	κ		OA	κ		OA	κ		OA	κ	
	81.15%	0.8296	30.172	80.06%	0.8137	33.847	81.82%	0.8123	31.118	84.51%	0.8119	30.633	65.98%	0.7602	31.647
Gravel	OA	κ		OA	κ		OA	κ		OA	κ		OA	κ	
	67.97%	0.7422		64.17%	0.7393		61.98%	0.7446		63.98%	0.7559		54.48%	0.7021	
Tiles	OA	κ		OA	κ		OA	κ		OA	κ		OA	κ	
	85.11%	0.8966		80.29%	0.8420		81.26%	0.8222		79.96%	0.8145		69.12%	0.7452	

8. Discussion and Conclusions

This research paper proposes a highly effective geographic object-based scene classification system for image *Remote Sensing* which employs a novel ZsL architecture. It introduces serious prerequisites for even more sophisticated pattern recognition systems without prior training.

To the best of our knowledge, it is the first time that such an algorithm has been presented in the literature. It facilitates learning of specialized intermediate representation extraction functions, under complex Deep Learning architectures. Additionally, it utilizes first and second order derivatives as a pre-training method for learning parameters which do not cause exploding or diminishing gradients. Finally, it avoids potential overfitting, while it significantly reduces computational costs and training time. It produces improved training stability, high overall performance and remarkable classification accuracy.

Likewise, the ensemble method used leads to much better prediction results, while providing generalization which is one of the key requirements in the field of machine learning [38]. At the same time, it reduces bias and variance and it eliminates overfitting, by implementing a robust model capable of responding to high complexity problems.

The proposed MAME-ZsL algorithm follows a heuristic hierarchical hyperparameter search methodology, using intermediate representations extracted from the employed neural network and avoiding other irrelevant ones. Using these elements, it discovers appropriate representations which can correctly classify unknown image samples. What should also be emphasized is the use of bootstrap sampling, which accurately addresses noisy scattered misclassification points which other spectral classification methods cannot handle.

The implementation of MAME-ZsL, utilizing the MAML++ algorithm, is based on the optimal use and combination of two highly efficient and fast learning processes (*Softmax* activation function and *AdaBound* algorithm) which create an integrated intelligent system. It is the first time which this hybrid approach has been introduced in the literature.

The successful choice of the *Softmax* (SFM) function instead of the *Sigmoid* (SIG) was based on the fact that it performs better on multi-classification problems, such as the one under consideration, and the sum of its probabilities equals to 1. On the other hand, the *Sigmoid* is used for binary classification tasks. Finally, in the case of SFM, high values have the highest probabilities, whereas in SIG this is not the case.

The use of the AdaBound algorithm offers high convergence speed compared to stochastic gradient descent models. Moreover, it exceeds the poor generalization ability of the adaptive approaches, as it has dynamic limits on the Learning Rate in order to obtain the highest accuracy for the dataset under consideration. Still, this network remarkably implements a GeoAI approach for large-scale geospatial data analysis which attempts to balance latency, throughput and fault-tolerance using ZsL. At the same time, it makes effective use of the intermediate representations of Deep Learning.

The proposed model avoids overfitting, decreases variance or bias, and it can fit unseen patterns, without reducing its performance.

The reliability of the proposed network has been proven in identifying scenes from Remote Sensing photographs. This suggests that it can be used in higher level geospatial data analysis processes, such as multi-sector classification, recognition and monitoring of specific patterns and sensors' data fusion. In addition, the performance of the proposed model was evaluated against state-of-the-art supervised Deep Learning models. It is worth mentioning that the proposed model was trained with instances only from seven classes, while the other models were trained with instances from all eleven classes. The presented numerical experiments demonstrate that the introduced approach produces remarkable results compared to theoretically superior models, providing convincing arguments regarding its classification efficiency.

Suggestions for the evolution and future improvements of this network should focus on comparison with other ZsL models. It is interesting to see the difference between ZsL, one-shot and five-shot learning methods, in terms of efficiency.

On the other hand, future research could focus on further optimization of the hyperparameters of the algorithms used in the proposed MAME-ZsL architecture. This may lead to an even more efficient, more accurate and faster classification process, either by using a heuristic approach or by employing a potential adjustment of the algorithm with spiking neural networks [44].

Additionally, it would be important to study the extension of this system by implementing more complex architectures with *Siamese neural networks* in parallel and distributed real time data stream environments [45].

Finally, an additional element which could be considered in the direction of future expansion, concerns the operation of the network by means of self-improvement and redefinition of its parameters automatically. It will thus be able to fully automate the process of extracting useful intermediate representations from Deep Learning techniques.

Author Contributions: Conceptualization, K.D. and L.I.; investigation, K.D.; methodology, K.D. and L.I.; software, K.D. and L.I.; validation, K.D. and L.I.; formal analysis L.I.; resources, K.D. and L.I.; data curation, K.D. and L.I.; writing—original draft preparation, K.D.; writing—review and editing L.I.; supervision, L.I. All authors have read and agreed to the published version of the manuscript.

Funding: This research received no external funding.

Conflicts of Interest: The authors declare no conflict of interest.

References

1. Petersson, H.; Gustafsson, D.; Bergstrom, D. Hyperspectral Image Analysis Using Deep Learning—A Review. In Proceedings of the 2016 Sixth International Conference on Image Processing Theory, Tools and Applications (IPTA), Oulu, Finland, 12–15 December 2016; pp. 1–6. [CrossRef]
2. Dixit, M.; Tiwari, A.; Pathak, H.; Astya, R. An Overview of Deep Learning Architectures, Libraries and Its Applications Areas. In Proceedings of the 2018 International Conference on Advances in Computing, Communication Control and Networking (ICACCCN), Greater Noida, India, 12–13 October 2018; pp. 293–297. [CrossRef]
3. Goodfellow, I.; Bengio, Y.; Courville, A. *Deep Learning*; MIT Press: Cambridge, MA, USA, 2016.
4. Vinyals, O.; Blundell, C.; Lillicrap, T.; Kavukcuoglu, K.; Wierstra, D. Matching networks for one shot learning. *arXiv* **2016**, arXiv:1606.04080.
5. Fu, Y.; Xiang, T.; Jiang, Y.; Xue, X.; Sigal, L.; Gong, S. Recent Advances in Zero-Shot Recognition: Toward Data-Efficient Understanding of Visual Content. *IEEE Signal Process. Mag.* **2018**, *35*, 112–125. [CrossRef]
6. Jamal, M.A.; Qi, G.; Shah, M. Task-agnostic meta-learning for few-shot learning. *arXiv* **2018**, arXiv:1805.07722.
7. Hochreiter, S.; Younger, A.S.; Conwell, P.R. Learning to Learn Using Gradient Descent. In Proceedings of the ICANN'01 International Conference, Vienna, Austria, 21–25 August 2001; pp. 87–94.
8. Lemke, C.; Budka, M.; Gabrys, B. Metalearning: A survey of trends and technologies. *Artif. Intell. Rev.* **2013**, *44*, 117–130. [CrossRef]
9. Finn, C.; Abbeel, P.; Levine, S. Alex Nichol and Joshua Achiam and John Schulman (2018). On First-Order Meta-Learning Algorithms. *arXiv* **2017**, arXiv:1803.02999.
10. Finn, C.; Levine, S. Meta-learning and universality: Deep representations and gradient descent can approximate any learning algorithm. *arXiv* **2017**, arXiv:1710.11622.
11. Vanschoren, J. Meta-Learning: A Survey. *arXiv* **2018**, arXiv:1810.03548.
12. Finn, C.; Abbeel, P.; Levine, S. Model-Agnostic Meta-Learning for Fast Adaptation of Deep Networks. *arXiv* **2017**, arXiv:1703.03400.
13. Nichol, A.; Achiam, J.; Schulman, J. On first-order meta-learning algorithms. *arXiv* **2018**, arXiv:1803.02999.
14. Antoniou, A.; Edwards, H.; Storkey, A. How to train your MAML. *arXiv* **2019**, arXiv:1810.09502.
15. Li, Z.; Zhou, F.; Chen, F.; Li, H. Meta-SGD: Learning to Learn Quickly for Few-Shot Learning. *arXiv* **2017**, arXiv:1707.09835.
16. Loshchilov, I.; Hutter, F. SGDR: Stochastic Gradient Descent with Warm Restarts. *arXiv* **2017**, arXiv:1608.03983.
17. Luo, L.; Xiong, Y.; Liu, Y.; Sun, X. Adaptive Gradient Methods with Dynamic Bound of Learning Rate. *arXiv* **2019**, arXiv:1902.09843.

18. Li, W.; Hsu, C.-Y. Automated terrain feature identification from remote sensing imagery: A deep learning approach. *Int. J. Geogr. Inf. Sci.* **2018**. [CrossRef]
19. Xie, Y.; Cai, J. A locally-constrained yolo framework for detecting small and densely-distributed building footprints. *Int. J. Geogr. Inf. Sci.* **2019**. [CrossRef]
20. Duan, W.; Chiang, Y.; Leyk, S.; Johannes, H.U.; Knoblock, C.A. Automatic alignment of contemporary vector data and georeferenced historical maps using reinforcement learning. *Int. J. Geogr. Inf. Sci.* **2019**, *34*, 2020. [CrossRef]
21. Guo, Z.; Feng, C.-C. Using multi-scale and hierarchical deep convolutional features for 3D semantic classification of tls point clouds. *Int. J. Geogr. Inf. Sci.* **2018**. [CrossRef]
22. Zhu, D. Spatial interpolation using conditional generative adversarial neural networks. *Int. J. Geogr. Inf. Sci.* **2019**. [CrossRef]
23. Polson, N.G.; Sokolov, V.O. Deep learning for short-term traffic flow prediction. *Transp. Res. Part C Emerg. Technol.* **2017**, *79*, 1–17. [CrossRef]
24. Fu, Y.; Hospedales, T.M.; Xiang, T.; Gong, S. Transductive Multi-View Zero-Shot Learning. *IEEE Trans. Pattern Anal. Mach. Intell.* **2015**, *37*, 2332–2345. [CrossRef]
25. Romera-Paredes, B.; Torr, P.H.S. An Embarrassingly Simple Approach to Zero-Shot Learning. In *Visual Attributes. Advances in Computer Vision and Pattern Recognition*; Feris, R., Lampert, C., Parikh, D., Eds.; Springer: Cham, Switzerland, 2017.
26. Sumbul, R.; Cinbis, G.; Aksoy, S. Fine-Grained Object Recognition and Zero-Shot Learning in Remote Sensing Imagery. *IEEE Trans. Geosci. Remote Sens.* **2018**, *56*, 770–779. [CrossRef]
27. Imamura, R.; Itasaka, T.; Okuda, M. Zero-Shot Hyperspectral Image Denoising with Separable Image Prior. In Proceedings of the IEEE International Conference on Computer Vision Workshops, Cambridge, MA, USA, 20–23 June 1995.
28. Hu, R.; Xiong, C.; Richard, S. Correction Networks: Meta-Learning for Zero-Shot Learning. 2018. Available online: https://openreview.net/forum?id=r1xurn0cKQ (accessed on 11 January 2020).
29. Zhang, T.X.; Gong, S. Learning a Deep Embedding Model for Zero-Shot Learning. In Proceedings of the 2017 IEEE Conference on Computer Vision and Pattern Recognition (CVPR), Honolulu, HI, USA, 21–26 July 2017; pp. 3010–3019. [CrossRef]
30. Gui, R. A generalized zero-shot learning framework for PolSAR land cover classification. *Remote Sens.* **2018**, *10*, 1307. [CrossRef]
31. Kotsiantis, S.; Pintelas, P. Combining Bagging and Boosting. *Int. J. Comput. Intell.* **2005**, *1*, 324–333.
32. Hyperspectral Remote Sensing Scenes. Available online: http://www.ehu.eus/ccwintco/index.php/Hyperspectral_Remote_Sensing_Scenes (accessed on 13 November 2019).
33. Kotsiantis, S.; Kanellopoulos, D.; Pintelas, P. Data Preprocessing for Supervised Learning. *Int. J. Comput. Sci.* **2006**, *1*, 111–117.
34. Livieris, I.; Pintelas, P. Performance Evaluation of Descent CG Methods for Neural Network Training. In Proceedings of the 9th Hellenic European Research on Computer Mathematics its Applications Conference HERCMA 2009, Athens, Greece, 24–26 September 2009.
35. Sim, J.; Wright, C.C. The Kappa Statistic in Reliability Studies: Use, Interpretation, and Sample Size Requirements. *Phys. Ther.* **2005**, *85*, 257–268. [CrossRef]
36. Agresti, A. *Categorical Data Analysis*; John Wiley & Sons: Hoboken, NJ, USA, 2002; p. 413. ISBN 978-0-471-36093-3.
37. He, K.; Zhang, X.; Ren, S.; Sun, J. Deep Residual Learning for Image Recognition. *arXiv* **2015**, arXiv:1512.03385.
38. Hu, W.; Huang, Y.; Wei, L.; Zhang, F.; Li, H. Deep convolutional neural networks for hyperspectral image classification. *J. Sens.* **2015**, *2015*, 258619. [CrossRef]
39. Chen, Y.; Jiang, H.; Li, C.; Jia, X.; Ghamisi, P. Deep feature extraction and classification of hyperspectral images based on CNN. *Trans. Geosci. Remote Sens.* **2016**, *54*, 6232–6251. [CrossRef]
40. Dosovitskiy, A.; Springenberg, J.T.; Brox, T. Learning to Generate Chairs, Tables and Cars with Convolutional Networks. *IEEE Trans. Pattern Anal. Mach. Intell.* **2017**, *39*, 692–705. [CrossRef]
41. Dosovitskiy, A.; Fischer, P.; Springenberg, J.T.; Riedmiller, M.; Brox, T. Discriminative unsupervised feature learning with exemplar convolutional neural networks. *IEEE Trans. Pattern Anal. Mach. Intell.* **2016**, *38*, 1734–1747. [CrossRef]

42. Mou, L.; Ghamisi, P.; Zhu, X.X. Unsupervised Spectral–Spatial Feature Learning via Deep Residual Conv–Deconv Network for Hyperspectral Image Classification. *IEEE Trans. Geosci. Remote Sens.* **2018**, *56*, 391–406. [CrossRef]
43. Demertzis, K.; Iliadis, L.; Anezakis, V.D. Commentary: Aedes albopictus and Aedes japonicus—Two invasive mosquito species with different temperature niches in Europe. *Front. Environ. Sci.* **2017**, *5*, 85. [CrossRef]
44. Demertzis, K.; Iliadis, L.; Anezakis, V.D. A Deep Spiking Machine-Hearing System for the Case of Invasive Fish Species. In Proceedings of the 2017 IEEE International Conference on Innovations in INtelligent SysTems and Applications (INISTA), Gdynia, Poland, 3–5 July 2017; pp. 23–28.
45. Demertzis, K.; Iliadis, L.; Anezakis, V.D. A Dynamic Ensemble Learning Framework for Data Stream Analysis and Real-Time Threat Detection. In *Artificial Neural Networks and Machine Learning—ICANN 2018*; Kůrková, V., Manolopoulos, Y., Hammer, B., Iliadis, L., Maglogiannis, I., Eds.; Lecture Notes in Computer Science; Springer: Cham, Switzerland, 2018; Volume 11139.

 © 2020 by the authors. Licensee MDPI, Basel, Switzerland. This article is an open access article distributed under the terms and conditions of the Creative Commons Attribution (CC BY) license (http://creativecommons.org/licenses/by/4.0/).

Article

Ensemble Learning of Hybrid Acoustic Features for Speech Emotion Recognition

Kudakwashe Zvarevashe and Oludayo Olugbara *

ICT and Society Research Group, South Africa Luban Workshop, Durban University of Technology, Durban 4001, South Africa; kudakwashe.zvarevashe@gmail.com
* Correspondence: oludayoo@dut.ac.za

Received: 15 January 2020; Accepted: 17 March 2020; Published: 22 March 2020

Abstract: Automatic recognition of emotion is important for facilitating seamless interactivity between a human being and intelligent robot towards the full realization of a smart society. The methods of signal processing and machine learning are widely applied to recognize human emotions based on features extracted from facial images, video files or speech signals. However, these features were not able to recognize the fear emotion with the same level of precision as other emotions. The authors propose the agglutination of prosodic and spectral features from a group of carefully selected features to realize hybrid acoustic features for improving the task of emotion recognition. Experiments were performed to test the effectiveness of the proposed features extracted from speech files of two public databases and used to train five popular ensemble learning algorithms. Results show that random decision forest ensemble learning of the proposed hybrid acoustic features is highly effective for speech emotion recognition.

Keywords: emotion recognition; ensemble algorithm; feature extraction; hybrid feature; machine learning; supervised learning

---

1. Introduction

Emotion plays an important role in the daily interpersonal interactions and is considered an essential skill for human communication [1]. It helps humans to understand the opinions of others by conveying feelings and giving feedback to people. Emotion has many useful benefits of affective computing and cognitive activities such as rational decision making, perception and learning [2]. It has opened up an exhilarating research agenda because constructing an intelligent robotic dialogue system that can recognize emotions and precisely respond in the manner of human conversation is presently arduous. The requirement of emotion recognition is steadily increasing with the pervasiveness of intelligent systems [3]. Huawei intelligent video surveillance systems, for instance, can support real-time tracking of a person in a distressed phase through emotion recognition. The capability to recognize human emotions is considered an essential future requirement of intelligent systems that are inherently supposed to interact with people to a certain degree of emotional intelligence [4]. The necessity to develop emotionally intelligent systems is exceptionally important for the modern society of the internet of things (IoT) because such systems have great impact on decision making, social communication and smart connectivity [5].

Practical applications of emotion recognition systems can be found in many domains such as audio/video surveillance [6], web-based learning, commercial applications [7], clinical studies, entertainment [8], banking [9], call centers [10], computer games [11] and psychiatric diagnosis [12]. In addition, other real applications include remote tracking of persons in a distressed phase, communication between human and robots, mining sentiments of sport fans and customer care services [13], where emotion is perpetually expressed. These numerous applications have led to the development of emotion recognition systems that use facial images, video files or speech signals [14]. In particular, speech signals carry emotional messages during their production [15] and have led to the development of intelligent systems habitually called speech emotion recognition systems [16]. There is an avalanche of intrinsic socioeconomic advantages that make speech signals a good source for affective computing. They are economically easier to acquire than other biological signals like electroencephalogram, electrooculography and electrocardiograms [17], which makes speech emotion recognition research attractive [18]. Machine learning algorithms extract a set of speech features with a variety of transformations to appositely classify emotions into different classes. However, the set of features that one chooses to train the selected learning algorithm is one of the most important tools for developing effective speech emotion recognition systems [3,19]. Research has suggested that features extracted from the speech signal have a great effect on the reliability of speech emotion recognition systems [3,20], but selecting an optimal set of features is challenging [21].

Speech emotion recognition is a difficult task because of several reasons such as an ambiguous definition of emotion [22] and the blurring of separation between different emotions [23]. Researchers are investigating different heterogeneous sources of features to improve the performance of speech emotion recognition systems. In [24], performances of Mel-frequency cepstral coefficient (MFCC), linear predictive cepstral coefficient (LPCC) and perceptual linear prediction (PLP) features were examined for recognizing speech emotion that achieved a maximum accuracy of 91.75% on an acted corpus with PLP features. This accuracy is relatively low when compared to the recognition accuracy of 95.20% obtained for a fusion of audio features based on a combination of MFCC and pitch for recognizing speech emotion [25]. Other researchers have tried to agglutinate different acoustic features with the optimism of boosting accuracy and precision rates of speech emotion recognition [25,26]. This technique has shown some improvement, nevertheless it has yielded low accuracy rates for the fear emotion in comparison with other emotions [25,26]. Semwal et al. [26] fused some acoustic features such as MFCC, energy, zero crossing rate (ZCR) and fundamental frequency that gave an accuracy of 77.00% for the fear emotion. Similarly, Sun et al. [27] used a deep neural network (DNN) to extract bottleneck features that achieved an accuracy of 62.50% for recognizing the fear emotion. The overarching objective of this study was to construct a set of hybrid acoustic features (HAFs) to improve the recognition of the fear emotion and other emotions from speech signal with high precision. This study has contributed to the understanding of the topical theme of speech emotion recognition in the following unique ways.

- The application of a specialized software to extract highly discriminating speech emotion feature representations from multiple sources such as prosodic and spectral to achieve an improved precision in emotion recognition.
- The agglutination of the extracted features using the specialized software to form a set of hybrid acoustic features that can recognize the fear emotion and other emotions from speech signal better than the state-of-the-art unified features.
- The comparison of the proposed set of hybrid acoustic features with other prominent features of the literature using popular machine learning algorithms to demonstrate through experiments, the effectiveness of our proposal over the others.

The content of this paper is succinctly organized as follows. Section 1 provides the introductory message, including the objective and contributions of the study. Section 2 discusses the related studies in chronological order. Section 3 designates the experimental databases and the study methods. The details of the results and concluding statements are given in Sections 4 and 5 respectively.

2. Related Studies

Speech emotion recognition research has been exhilarating for a long time and several papers have presented different ways of developing systems for recognizing human emotions. The authors in [3] presented a majority voting technique (MVT) for detecting speech emotion using fast correlation based feature (FCBF) and Fisher score algorithms for feature selection. They extracted 16 low-level features and tested their methods over several machine learning algorithms, including artificial neural network (ANN), classification and regression tree (CART), support vector machine (SVM) and K-nearest neighbor (KNN) on berlin emotion speech database (Emo-DB). Kerkeni et al. [11] proposed a speech emotion recognition method that extracted MFCC with modulation spectral (MS) features and used recurrent neural network (RNN) learning algorithm to classify seven emotions from Emo-DB and Spanish database. The authors in [15] proposed a speech emotion recognition model where glottis was used for compensation of glottal features. They extracted features of glottal compensation to zero crossings with maximal teager (GCZCMT) energy operator using Taiyuan University of technology (TYUT) speech database and Emo-DB.

Evaluation of feature selection algorithms on a combination of linear predictive coefficient (LPC), MFCC and prosodic features with three different multiclass learning algorithms to detect speech emotion was discussed in [19]. Luengo et al. [20] used 324 spectral and 54 prosody features combined with five voice quality features to test their proposed speech emotion recognition method on the Surrey audio-visual expressed emotion (SAVEE) database after applying the minimal redundancy maximal relevance (mRMR) to reduce less discriminating features. In [28], a method for recognizing emotions in an audio conversation based on speech and text was proposed and tested on the SemEval-2007 database using SVM. Liu et al. [29] used the extreme learning machine (ELM) method for feature selection that was applied to 938 features based on a combination of spectral and prosodic features from Emo-DB for speech emotion recognition. The authors in [30] extracted 988 spectral and prosodic features from three different databases using the OpenSmile toolkit with SVM for speech emotion recognition. Stuhlsatz et al. [31] introduced the generalized discriminant analysis (GerDA) based on DNN to recognize emotions from speech using the OpenEar specialized software to extract 6552 acoustic features based on 39 functional of 56 acoustic low-level descriptor (LLD) from Emo-DB and speech under simulated and actual stress (SUSAS) database. Zhang et al. [32] applied the cooperative learning method to recognize speech emotion from FAU Aibo and SUSAS databases using GCZCMT based acoustic features.

In [33], Hu moments based weighted spectral features (HuWSF) were extracted from Emo-DB, SAVEE and Chinese academy of sciences - institute of automation (CASIA) databases to classify emotions from speech using SVM. The authors used HuWSF and multicluster feature selection (MCFS) algorithm to reduce feature dimensionality. In [34], extended Geneva minimalistic acoustic parameter set (eGeMAPS) features were used to classify speech emotion, gender and age from the Ryerson audio-visual database of emotional speech and song (RAVDESS) using the multiple layer perceptron (MLP) neural network. Pérez-Espinosa et al. [35] analyzed 6920 acoustic features from the interactive emotion dyadic motion capture (IEMOCAP) database to discover that features based on groups of MFCC, LPC and cochleagrams are important for estimating valence, activation and dominance emotions in speech respectively. The authors in [36] proposed a DNN architecture for extracting informative feature representatives from heterogeneous acoustic feature groups that may contain redundant and unrelated information. The architecture was tested by training the fusion network to jointly learn highly discriminating acoustic feature representations from the IEMOCAP database for speech emotion recognition using SVM to obtain an overall accuracy of 64.0%. In [37], speech features based on MFCC and facial on maximally stable extremal region (MSER) were combined to recognize human emotions through a systematic study on Indian face database (IFD) and Emo-DB.

Narendra and Alku [38] proposed a new dysarthric speech classification method from a coded telephone speech using glottal features with a DNN based glottal inverse filtering method. They considered two sets of glottal features based on time and frequency domain parameters plus parameters based on principal component analysis (PCA). Their results showed that a combination of glottal and acoustic features resulted in an improved classification after applying PCA. In [39], four pitch and spectral energy features were combined with two prosodic features to distinguish two high activation states of angry and happy plus low activation states of sadness and boredom for speech emotion recognition using SVM with Emo-DB. Alshamsi et al. [40] proposed a smart phone method for automated facial expression and speech emotion recognition using SVM with MFCC features extracted from SAVEE database.

Li and Akagi [41] presented a method for recognizing emotions expressed in a multilingual speech using the Fujitsu, Emo-DB, CASIA and SAVEE databases. The highest weighted average precision was obtained after performing speaker normalization and feature selection. The authors in [42] used MFCC related features for recognizing speech emotion based on an improved brain emotional learning (BEL) model inspired by the emotional processing mechanism of the limbic system in human brain. They tested their method on CASIA, SAVEE and FAU Aibo databases using linear discriminant analysis (LDA) and PCA for dimensionality reduction to obtain the highest average accuracy of 90.28% on CASIA database. Mao et al. [43] proposed the emotion-discriminative and domain-invariant feature learning method (EDFLM) for recognizing speech emotion. In their method, both domain divergence and emotion discrimination were considered to learn emotion-discriminative and domain-invariant features using emotion label and domain label constraints. Wang et al. [44] extracted MFCC, Fourier parameters, fundamental frequency, energy and ZCR from three different databases of Emo-DB, Chinese elderly emotion database (EESDB) and CASIA for recognizing speech emotion. Muthusamy et al. [45] extracted a total of 120 wavelet packet energy and entropy features from speech signals and glottal waveforms from Emo-DB, SAVEE and Sahand emotional speech (SES) databases for speech emotion recognition. The extracted features were enhanced using the Gaussian mixture model (GMM) with ELM as the learning algorithm.

Zhu et al. [46] used a combination of acoustic features based on MFCC, pitch, formant, short-term ZCR and short-term energy to recognize speech emotion. They extracted the most discriminating features and performed classification using the deep belief network (DBN) with SVM. Their results showed an average accuracy of 95.8% on the CASIA database across six emotions, which is an improvement when compared to other related studies. Álvarez et al. [47] proposed a classifier subset selection (CSS) for the stacked generalization to recognize speech emotion. They used the estimation of distribution algorithm (EDA) to select optimal features from a collection of features that included eGeMAPS and SVM for classification that achieved an average accuracy of 82.45% on Emo-Db. Bhavan et al. [48] used a combination of MFCCs, spectral centroids and MFCC derivatives of spectral features with a bagged ensemble algorithm based on Gaussian kernel SVM for recognizing speech emotion that achieved an accuracy of 92.45% on Emo-DB. Shegokar et al. [49] proposed the use of continuous wavelet transform (CWT) with prosodic features to recognize speech emotion. They used PCA for feature transformation with quadratic kernel SVM as a classification algorithm that achieved an average accuracy of 60.1% on the RAVDESS database. Kerkeni et al. [50] proposed a model for recognizing speech emotion using empirical mode decomposition (EMD) based on optimal features that included the reconstructed signal based on Mel frequency cepstral coefficient (SMFCC), energy cepstral coefficient (ECC), modulation frequency feature (MFF), modulation spectral (MS) and frequency weighted energy cepstral coefficient (FECC). They achieved an average accuracy of 91.16% on the Spanish database using the RNN algorithm for classification.

Emotion recognition is still a great challenge because of several reasons as previously alluded in the introductory message. Further reasons include the existence of a gap between acoustic features and human emotions [31,36] and the non-existence of a solid theoretical foundation relating the characteristics of voice to the emotions of a speaker [20]. These intrinsic challenges have led to the disagreement in the literature on which features are best for speech emotion recognition [20,36]. The trend in the literature shows that a combination of heterogeneous acoustic features is promising for speech emotion recognition [20,29,30,36,39], but how to effectively unify the different features is highly challenging [21,36]. The importance of selecting relevant features to improve the reliability of speech emotion recognition systems is strongly emphasized in the literature [3,20,29]. Researchers frequently apply specialized software such as OpenEar [31,32,35], OpenSmile [30,32,34,36] and Praat [35] for extraction, selection and unification of speech features from heterogeneous sources to ease the intricacy inherent in the processes. Moreover, the review of numerous literature has revealed that different learning algorithms are trained and validated with the specific features extracted from public databases for speech emotion recognition. Table 1 shows the result of the comparative analysis of our method with the related methods in terms of the experimental database, type of recording, number of emotions, feature set, classification method and maximum percentage accuracy results obtained. It is conspicuous from the comparative analysis that our emotion recognition method is highly promising because it has achieved the highest average accuracy of 99.55% across two dissimilar public experimental databases.

Table 1. A comparison of our method with related methods.

Reference	Database	Type of Recording	Number of Emotions	Feature Set	Classification Method	Result (%)
[3]	Emo-DB	Acted	Angry, Happy, Neutral, Sad	Energy + MFCC + ZCR + voicing probability + fundamental frequency	FCBF + MVT	84.19
[11]	Spanish	Acted	Anger, Disgust, Fear, Neutral, Surprise, Sadness, Joy	MFCC + MS	RNN classifier	90.05
[15]	Emo-DB	Acted	Angry, Happy, Neutral	GCZCMT	SVM	84.45
[19]	Emo-DB	Acted	Anger, Happiness, Neutral, Sadness	Prosodic + sub-band + MFCC + LPC	SFS algorithm + SVM	83.00
[28]	eNTERFACE'05	Elicited	Disgust, Surprise, Happy, Anger, Sad, Fear,	Pitch, energy, formants, intensity and ZCR + text	SVM	90.00
[29]	CASIA	Acted	Neutral, Happy, Sadness, Fear, Angry, Surprise	Prosodic + quality characteristics + MFCC	correlation analysis + Fisher + ELM decision tree	89.60
[36]	IEMOCAP	Acted	Angry, Happy, Neutral, Sad	IS10 + MFCCs + eGemaps + SoundNet + VGGish	SVM	64.00
[39]	Emo-DB	Acted	Anger, Boredom, Happy, Neutral, Sadness	Prosodic features + paralinguistic features	SVM	94.90
[42]	CASIA	Acted	Surprise, Happy, Sad, Angry, Fear, Neutral	MFCC	GA-BEL + PCA + LDA	90.28
[46]	CASIA	Acted	Angry, Fear, Happy, Neutral, Surprise, Sad	MFCC, pitch, formant, ZCR and short-term energy	SVM + DBN	95.80
[47]	Emo-DB	Acted	Sadness, Fear, Joy, Anger, Surprise, Disgust, Neutral	eGeMAPS	CSS stacking system +SVM	82.45
[48]	Emo-DB	Acted	Anger, Happiness, Sadness, Boredom, Neutral, Disgust, Fear,	MFCCs	bagged ensemble of SVMs	92.45
[49]	RAVDESS	Acted	Neutral, Surprise, Happy, Angry, Calm, Sad, Fearful, Disgust	CWT, prosodic coefficients	SVM	60.10
[50]	Spanish	Acted	Anger, Joy, Disgust, Neutral, Surprise, Fear, Sadness	SMFCC, ECC, MFF, MS and EFCC	RNN	91.16
Proposed model	RAVDESS/SAVEE	Acted	Angry, Sad, Happy, Disgust, Calm, Fear, Neutral, Surprise	Prosodic + spectral	RDF ensemble	99.55

3. Materials and Methods

Materials used for this study included speech emotion multimodal databases of RAVDESS [51] and SAVEE [52]. The databases, because of their popularity were chosen to test for the effectiveness of the proposed HAF against two other sets of features. The study method followed three phases of feature extraction, feature selection and classification, which were subsequently described. The method was heralded by the feature extraction process that involves the abstraction of prosodic and spectral features from the raw audio files of RAVDESS and SAVEE databases. This phase was subsequently followed by the feature selection process that involved the filtration, collection and agglutination of features that have high discriminating power of recognizing emotions in human speech. Classification was the last phase that involved the application of the selected learning algorithm to recognize human emotions and a comparative analysis of experimental results of classification. In the classification phase, several experiments were carried out on a computer with an i7 2.3GHz processor and 8 GB of random access memory (RAM). The purpose of the experiments was to apply standard metrics of accuracy, precision, recall and F1-score to evaluate the effectiveness of the proposed HAF with respect to a given database and ensemble learning algorithm.

3.1. Databases

The speech emotion database is an important precursor for analyzing speech and recognizing emotions. The database provides speech data for training and testing the effectiveness of emotion recognition algorithms [15]. RAVDESS and SAVEE are two public speech emotion databases used in this study for experiments to verify the effectiveness of the proposed HAF for emotion recognition. The two selected experimental databases were fleetingly discussed in this subsection of the paper.

3.1.1. RAVDESS

RAVDESS is a gender balanced set of validated speeches and songs that consists of eight emotions of 24 professional actors speaking similar statements in a North American accent. It is a multiclass database of angry, calm, disgust, fear, happy, neutral, sad and surprise emotions with 1432 American English utterances. Each of the 24 recorded vocal utterances comprises of three formats, which are audio-only (16bit, 48kHz .wav), audio-video (720p H.264, AAC 48kHz, .mp4) and video-only (no sound). The audio-only files were used across all the eight emotions because this study concerns speech emotion recognition. Figure 1 shows that angry, calm, disgust, fear, happy and sad emotion classes constituted 192 audio files each. The surprise emotion had 184 files and the neutral emotion had the lowest number of audio files of 96.

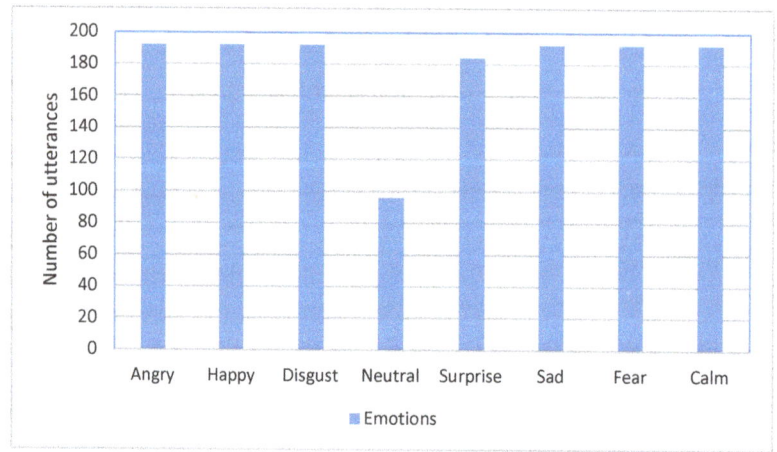

Figure 1. The Ryerson audio-visual database of emotional speech and song (RAVDESS).

3.1.2. SAVEE

SAVEE is a speech emotion database that consists of recordings from four male actors in seven different emotion classes. The database comprises of a total of 480 British English utterances, which is quite different from the North American accent used in the RAVDESS database. These vocal utterances were processed and labeled in a standard media laboratory with high quality audio-visual equipment. The recorded vocal utterances were classified into seven classes of angry, disgust, fear, happy, neutral, sad and surprise emotional expressions. The neutral emotion constituted 120 audio files, while all the other remaining emotions comprised of 60 audio files each as illustrated in Figure 2.

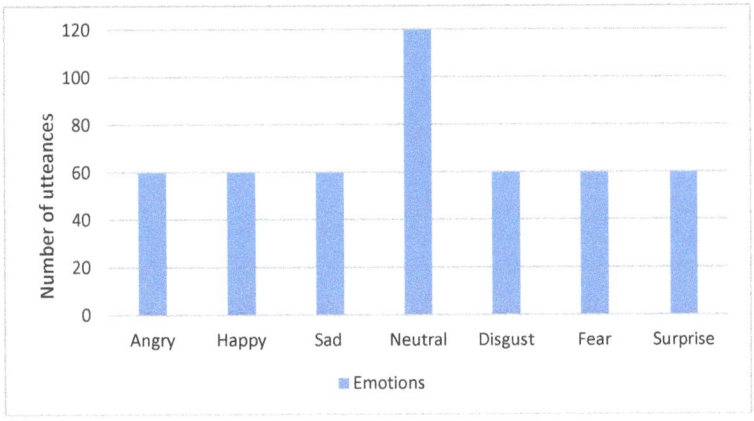

Figure 2. The Surrey audio-visual expressed emotion (SAVEE).

3.2. Feature Extraction

The speech signal carries a large number of useful information that reflects emotion characteristics such as gender, age, stuttering and identity of the speaker. Feature extraction is an important mechanism in audio processing to capture the information in a speech signal and most of the related studies have emphasized on the extraction of low-level acoustic features for speech recognition. Feature representation plays a prominent role in distinguishing the speech of speakers from each other. Since no universal consensus on the best features for speech recognition, certain authors have mentioned that prosody carries most of the useful information about emotions and can create a suitable feature representation [20]. Moreover, spectral features describe the properties of a speech signal in the frequency domain to identify important aspects of speech and manifest the correlation between vocal movements and changes of channel shape [35,42]. Prosodic and spectral acoustic features were suggested as the most important characteristics of speech because of the improved recognition results of their integration [25,42,53,54]. The blend of prosodic and spectral features is capable of enhancing the performance of emotion recognition, giving better recognition accuracy and a lesser number of emotions in a gender dependent model than many existing systems with the same number of emotions [25]. Banse et al. [54] examined vocal cues for 14 emotions using a combination of prosodic speech features and spectral information in a voiced segment and unvoiced segment to achieve impressive results.

The present authors have applied the selective approach [35] to extract prosodic and spectral features, taking cognizance of features that could be useful for improving emotion recognition. These are features that have produced positive results in the related studies and those from the other comparable tasks. We carefully applied the specialized jAudio software [55] to construct MFCC1, MFCC2 and HAF as three sets of features representing several voice aspects. MFCC1 is the set of MFCC features inspired by the applications of MFCC features [24,40,42]. MFCC2 is the set of features based on

MFCC, energy, ZCR and fundamental frequency as inspired by the fusion of MFCC with other acoustic features [11,25,26,44,46]. The HAF is the proposed set of hybrid acoustic features of prosodic and spectral carefully selected based on the interesting results in the literature [11,20,24–26,29,30,39,40,42,44,46,48].

The feature extraction process was aided by the application of a specialized jAudio to effectively manage the inherent complexity in the selection process. The jAudio software is a digital signal processing library of advanced audio feature extraction algorithms designed to eliminate effort duplication in manually calculating emotion features from an audio signal. It provides a unique method of handling multidimensional features, dependency and allows for iterative development of new features from the existing ones through the meta-feature template. Moreover, it simplifies the intrinsic stiffness of the existing feature extraction methods by allowing low-level features to be fused to build increasingly high-level musically meaningful features. In addition, it provides audio samples as simple vectors of features and offers aggregation functions that aggregate a sequence of separate feature sets into a single feature vector.

3.2.1. Prosodic Features

Prosodic features are acoustic features prominently used in emotion recognition and speech signal processing because they carry essential paralinguistic information. This information complements a message with an intention that can paint a flawless picture about attitude or emotion [35,56]. In addition, prosodic features are considered as suprasegmental information because they help in defining and structuring the flow of speech [35]. Prosodic continuous speech features such as pitch and energy convey much content of emotions in speech [57] and are important for delivering the emotional cues of the speakers. These features include formant, timing and articulation features and they characterize the perceptual properties of speech typically used by human beings to perform different speech tasks [57].

The present authors have included three important prosodic features of energy, fundamental frequency and ZCR. Signal energy models the voice intensity, volume or loudness and reflects the pause and ascent of the voice signal. It is often associated with the human respiratory system and is one of the most important characteristics of human aural perception. The logarithm function is often used to reflect minor changes of energy because the energy of an audio signal is influenced by the recording conditions. Fundamental frequency provides tonal plus rhythmic characteristics of a speech and carries useful information about the speaker. ZCR determines the information about the number of times a signal waveform crosses the zero amplitude line because of a transition from a positive/negative value to a negative/positive value in a given time. It is suitable for detecting voice activity, end point, voiced sound segment, unvoiced sound segment, silent segment and approximating the measure of noisiness in speech [58]. ZCR is an acoustic feature that has been classified as a prosodic feature [49,59,60]. In particular, energy and pitch were declared prosody features with a low frequency domain while ZCR and formants are high frequency features [60].

3.2.2. Spectral Features

Spectral features of this study include timbral features that have been successful in music recognition [61]. Timbral features define the quality of a sound [62] and they are a complete opposite of most general features like pitch and intensity. It has been revealed that a strong relationship exists between voice quality and emotional content in a speech [58]. Siedenburg et al. [63] considered spectral features as significant in distinguishing between classes of speech and music. They claimed that the temporal evolution of the spectrum of audio signals mainly accounts for the timbral perception. Timbral features of a sound help in differentiating between sounds that have the same pitch and loudness and they assist in the classification of audio samples with similar timbral features into unique classes [61]. The application of spectral timbral features like spectral centroid, spectral roll-off point, spectral flux, time domain zero crossings and root mean squared energy amongst others has been demonstrated for speech analysis [64].

The authors have included eight spectral features of MFCC, spectral roll-off point, spectral flux, spectral centroid, spectral compactness, fast Fourier transforms, spectral variability and LPCC. MFCC is one of the most widely used feature extraction methods for speech analysis because of its computational simplicity, superior ability of distinction and high robustness to noise [42]. It has been successfully applied to discriminate phonemes and can determine what is said and how it is said [35]. The method is based on the human hearing system that provides a natural and real reference for speech recognition [46]. In addition, it is based on the sense that human ears perceive sound waves of different frequencies in a non-linear mode [42]. Spectral roll-off point defines a frequency below which at least 85% of the total spectral energy are present and provides a measure of spectral shape [65]. Spectral flux characterizes the dynamic variation of spectral information, is related to perception of music rhythm and it captures spectrum difference between two adjacent frames [66].

Spectral centroid describes the center of gravity of the magnitude spectrum of short-time Fourier transform (STFT). It is a spectral moment that is helpful in modeling sharpness or brightness of sound [67]. Spectral compactness is obtained by comparing the components in the magnitude spectrum of a frame and magnitude spectrum of neighboring frames [68]. Fast Fourier transform (FFT) is a useful method of analyzing the frequency spectrum of a speech signal and features based on the FTT algorithm have the strongest frequency component in Hertz [69,70]. Spectral variability is a measure of variance of the magnitude spectrum of a signal and is attributed to different sources, including phonetic content, speaker, channel, coarticulation and context [71]. Spectral compactness is an audio feature that measures the noisiness of a speech signal and is closely related to the spectral smoothness of speech signal [72]. LPCCs are spectral features obtained from the envelope of LPC and are computed from sample points of a speech waveform. They have low vulnerability to noise, yielded a lower error rate in comparison to LPC features and can discriminate between different emotions [73]. LPCC features are robust, but they are not based on an auditory perceptual frequency scale like MFCC [74].

3.3. Feature Selection

Feature selection is a process of filtering insignificant features to create a subset of the most discriminating features in the input data. It reduces the cause of dimensionality and improves the success of a recognition algorithm. Based on the literature review, we observed that most methods used for feature selection are computationally expensive because of the enormous processing time involved. Moreover, accuracy and precision values of some of these methods are reduced because of redundancy, less discriminating features and poor accuracy values in certain emotions such as the fear emotion. The authors have selected 133 MFCC spectral features for the MFCC1 and 90 features based on MFCC, ZCR, energy and fundamental frequency for the MFCC2. In a nutshell, the effectiveness of the HAF was tested against one homogenous set of acoustic features (MFCC1) and another hybrid set of acoustic features (MFCC2).

Table 2 presents the list of prosodic and spectral features of this study in terms of group, type and number of features with their statistical characteristics obtained with the aid of jAudio that generated a csv file with numerical values for each feature together with the respective emotion class. The literature inspired brute force approach was then explored to select non-redundant features that are likely to help improve emotion recognition performance. This implies inspecting through a series of experiments for possible combinations of features to discover the right features that define HAF. This approach is although laborious to require the help of a stochastic optimization technique, it has yielded excellent performance results in this study.

Table 2. List of the selected prosodic and spectral features.

Group	Type	Number
Prosodic		
Energy	Logarithm of Energy	10
Pitch	Fundamental Frequency	70
Times	Zero Crossing Rate	24
Spectral		
Cepstral	MFCC	133
Shape	Spectral Roll-off Point	12
Amplitude	Spectral Flux	12
Moment	Spectral Centroid	22
Audio	Spectral Compactness	10
Frequency	Fast Fourier Transform	9
Signature	Spectral Variability	21
Envelope	LPCC	81
Total		404

3.4. Classification

Emotion classification is aimed at obtaining an emotional state for the input feature vector using a machine learning algorithm. In this study, classification experiments were conducted using ensemble learning algorithms to verify the effectiveness of the proposed HAF. Ensemble learning is considered a state-of-the-art approach for solving different machine learning problems by combining the predictions of multiple base learners [75]. It improves the overall predictive performance, decreases the risk of obtaining a local minimum and provides a better fit to the data space by combining the predictions of several weak learners into a strong learning algorithm. In this study, we applied bagging and boosting machine learning because they are widely used effective approaches for constructing ensemble learning algorithms [76–79]. Bagging is a technique that utilizes bootstrap sampling to reduce the variance of a decision tree and improve the accuracy of a learning algorithm by creating a collection of learning algorithms that are learned in parallel [79–81]. It performs random sampling with replacement over a simple averaging of all the predictions from different decision trees to give a more robust result than a single decision tree. Boosting creates a collection of learning algorithms that are learned sequentially with early learners to fit decision trees to the data and analyze data for errors. It performs random sampling with replacement over a weighted average and reduces classification bias and variance of a decision tree [79]. One of the bagging ensemble algorithms investigated in this study was random decision forest (RDF), which is an extension over bagging that is popularly called random forest [75,82]. RDF can be constituted by making use of bagging based on the CART approach to raise trees [83]. The other bagging ensemble learning algorithms investigated in this study were Bagging with SVM as the base learner (BSVM) and bagging with the multilayer perceptron neural network as the base learner (BMLP). The boosting ensemble algorithms investigated in this study were the gradient boosting machine (GBM), which extends boosting by combining the gradient descent optimization algorithm with boosting technique [75,82,84], and AdaBoost with CART as the base learner (ABC), which is one of the most widely used boosting algorithm to reduce sensitivity to class label noise [79,85]. AdaBoost is an iterative learning algorithm for constructing a strong classifier by enhancing weak classification algorithms and it can improve data classification ability by reducing both bias and variance through continuous learning [81].

The standard metrics of accuracy, precision, recall and F1-score have been engaged to measure the performances of the learning algorithms with respect to a particular set of features. The accuracy of a classification algorithm is often judged as one of the most intuitive performance measures. It is the ratio of the number of instances correctly recognized to the total number of instances. Precision is the ratio of the number of positive instances correctly recognized to the total number of positive instances. Recall is the ratio of the number of positive instances correctly recognized to the number of all the instances of the actual class [86]. F1-score constitutes a harmonic mean of precision and recall [87]. Moreover, the training times of the learning algorithms are also compared to analyze the computational complexities of training different ensemble learning classifiers. The flow chart for different configurations of the methods of experimentation is illustrated in Figure 3.

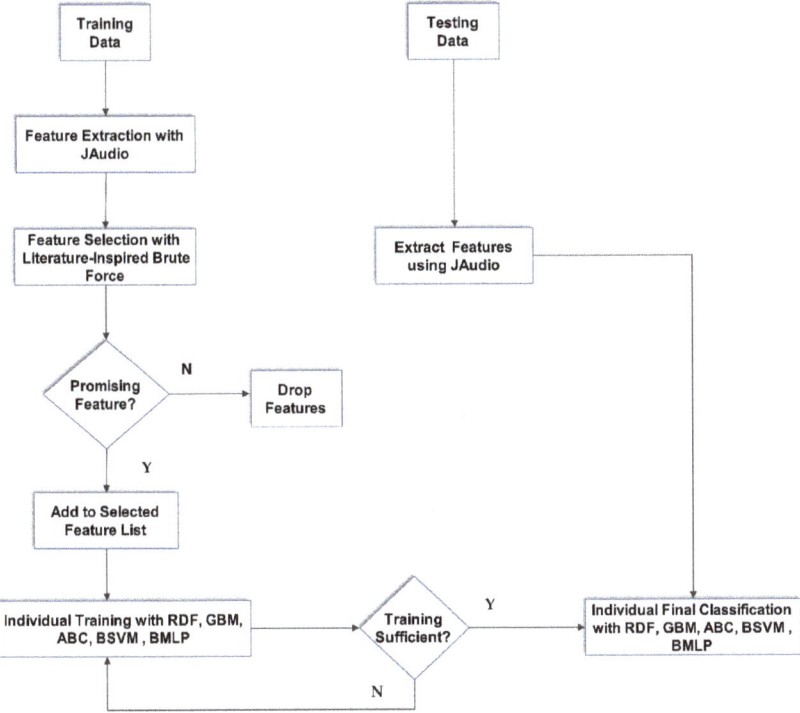

Figure 3. Flowchart for different configurations of emotion recognition methods.

4. Results and Discussion

The present authors have extracted the set of 404 HAFs from RAVDESS and SAVEE experimental databases. The features were used to train five famous ensemble learning algorithms in order to determine their effectiveness using four standard performance metrics. The performance metrics were selected because the experimental databases were not equally distributed. Data were subjected to 10 fold cross-validation and divided into training and testing groups as the norm permits. Gaussian noise was added to the training data to provide a regularizing effect, reduce overfitting, increase resilience and improve generalization performance of the ensemble learning algorithms.

Table 3 shows the result of the CPU computational time of training an individual ensemble learning algorithm with three different sets of features across two experimental databases. The training time analysis shows that RDF spent the least amount of training time when compared to other ensemble learning algorithms. The least training time was recorded with respect to MFCC1 and MFCC2 irrespective of databases. However, the training time of RDF (0.43) was slightly higher than that of BSVM (0.39) with respect to SAVEE HAF, but it claimed superiority with respect to the

RAVDESS HAF. The reason for this inconsistent result might be because of the smaller data size (480) of SAVEE when compared to the size (1432) of RAVDESS. The experimental results of a study that was aimed at fully assessing the predictive performances of SVM and SVM ensembles over small and large scale datasets of breast cancer show that SVM ensemble based on BSVM method can be the better choice for a small scale dataset [78]. Moreover, it can be observed that GBM and BMLP were generally time consuming when compared to the other ensemble learning algorithms because they took 1172.3 milliseconds to fit MFCC2 RAVDESS and 1319.39 milliseconds to fit MFCC2 SAVEE features respectively.

Table 3. Processing time of learning algorithms trained with Mel-frequency cepstral coefficient 1 (MFCC1), Mel-frequency cepstral coefficient 2 (MFCC2) and a hybrid acoustic feature (HAF).

Classifier	SAVEE Feature			RAVDESS Feature		
	MFCC1	MFCC2	HAF	MFCC1	MFCC2	HAF
RDF	0.03	0.38	0.43	0.052	1.62	0.06
GBM	2.30	579.40	343.10	5.90	1172.30	6.90
ABC	0.47	22.16	1.38	2.59	2.79	1.89
BSVM	0.20	49.79	0.39	0.42	211.54	0.48
BMLP	3.13	1319.39	952.46	76.00	231.16	7.18

The experimental results of the overall recognition performance obtained are given in Table 4. The percentage average recognition performance rates differ depending on critical factors such as the type of database, the accent of the speakers, the way data were collected, the type of features and the learning algorithm used. In particular, RDF consistently recorded the highest average accuracy, while BMLP consistently recorded the lowest average accuracy across the sets of features, irrespective of databases. These results indicate that RDF gave the highest performance because the computed values of the precision, recall, F1-score and accuracy of the ensemble learning algorithm are generally impressive when compared to other ensemble learning algorithms. However, average recognition accuracies of 61.5% and 66.7% obtained by RDF for the SAVEE MFCC1 and MFCC2 were respectively low, but the values were better than that of BMLP (55.4% and 60.7%) for SAVEE MFCC1 and MFCC2 respectively. In general, an improvement can be observed when a combination of features (MFCC2) was used in an experiment drawing inspiration from the work done in [46], where MFCC features were combined with pitch and ZCR features. In addition, we borrowed the same line of thought from Sarker and Alam [3] and Bhaskar et al. [28] where some of the above mentioned features were combined to form hybrid features.

Table 4. Percentage average precision, recall, F1-score and accuracy with confidence intervals of learning algorithms trained with MFCC1, MFCC2 and HAF.

Classifier/Measure	SAVEE Feature			RAVDESS Feature		
	MFCC1	MFCC2	HAF	MFCC1	MFCC2	HAF
RDF						
Precision	63.4 (±0.044)	78.1 (±0.037)	99.1(±0.009)	90.1 (±0.016)	93.1 (±0.013)	99.6 (±0.003)
Recall	72.1 (±0.041)	77.4 (±0.038)	99.1(±0.009)	88.9 (±0.016)	94.0 (±0.012)	99.5 (±0.004)
F1-score	63.6 (±0.043)	76.6 (±0.039)	99.1(±0.009)	88.5 (±0.017)	92.5 (±0.014)	99.8 (±0.002)
Accuracy	61.5 (±0.044)	66.7 (±0.043)	99.3 (±0.008)	90.7 (±0.015)	93.7 (±0.013)	99.8 (±0.002)
GBM						
Precision	64.0 (±0.043)	75.6 (±0.039)	99.4 (±0.007)	85.0 (±0.019)	86.5 (±0.018)	95.3(±0.011)
Recall	66.6 (±0.043)	74.6 (±0.039)	99.1 (±0.009)	82.9 (±0.020)	88.1 (±0.017)	94.8(±0.012)
F1-score	62.6 (±0.044)	72.7 (±0.040)	99.3 (±0.008)	82.6 (±0.020)	85.9 (±0.018)	96.8(±0.009)
Accuracy	61.5 (±0.044)	65.5 (±0.043)	99.3 (±0.008)	85.4(±0.018)	86.7(±0.018)	92.6(±0.014)
ABC						
Precision	62.9 (±0.044)	73.4 (±0.040)	99.1 (±0.009)	81.3 (±0.020)	85.4 (±0.018)	94.3(±0.012)
Recall	64.1 (±0.044)	73.7 (±0.040)	99.4 (±0.007)	84.0 (±0.019)	87.5 (±0.017)	94.5(±0.012)
F1-score	63.1 (±0.044)	74.1 (±0.040)	97.9 (±0.013)	82.6 (±0.020)	84.5 (±0.019)	95.9(±0.010)
Accuracy	58.0 (±0.044)	62.8 (±0.043)	98.0 (±0.008)	83.0 (±0.020)	85.2 (±0.018)	92.0 (±0.014)
BSVM						
Precision	61.4 (±0.044)	71.4 (±0.041)	98.7 (±0.010)	81.0 (±0.020)	84.1 (±0.019)	91.1(±0.015)
Recall	61.0 (±0.044)	72.6 (±0.041)	99.0 (±0.009)	82.5 (±0.020)	82.3 (±0.020)	92.0(±0.014)
F1-score	61.9 (±0.044)	72.0 (±0.041)	99.3 (±0.008)	80.5 (±0.021)	83.5 (±0.019)	90.3(±0.015)
Accuracy	56.0 (±0.045)	61.5 (±0.044)	96.0 (±0.013)	82.7 (±0.020)	84.5 (±0.019)	91.8 (±0.014)
BMLP						
Precision	59.4 (±0.044)	69.0 (±0.042)	97.9 (±0.013)	78.8 (±0.021)	77.5 (±0.022)	93.0 (±0.013)
Recall	60.1 (±0.044)	72.0 (±0.041)	98.4 (±0.011)	75.8 (±0.022)	80.3 (±0.021)	88.0(±0.017)
F1-score	59.6 (±0.044)	70.3 (±0.041)	98.1 (±0.012)	74.4(±0.023)	78.3 (±0.021)	89.0 (±0.016)
Accuracy	55.4 (±0.045)	60.7 (±0.044)	94.6 (±0.002)	75.6 (±0.022)	79.3 (±0.021)	91.3 (±0.015)

The results in Table 4 can be used to demonstrate the effectiveness of the proposed HAF. The values of the precision, recall, F1-score and accuracy performance measures can be seen to be generally high for the five ensemble learning algorithms using the HAF irrespective of databases. Specifically, RDF and GBM concomitantly recorded the highest accuracy of 99.3% for the SAVEE HAF, but RDF claims superiority with 99.8% accuracy when compared to the 92.6% accuracy for the HAF RAVDESS. Consequently, the results of RDF learning of the proposed HAF were exceptionally inspiring with an average accuracy of 99.55% across the two databases. The ranking of the investigated ensemble learning algorithms in terms of percentage average accuracy computed for HAF across the databases is RDF (99.55%), GBM (95.95%), ABC (95.00%), BSVM (93.90%) and BMLP (92.95%). It can be inferred that application of ensemble learning of HAF is highly promising for recognizing emotions in human speech. In general, high accuracy values computed by all learning algorithms across databases indicate the effectiveness of the proposed HAF for speech emotion recognition. Moreover, the confidence factors show that the true classification accuracies of all the ensemble learning algorithms investigated lie in the range between 0.002% and 0.045% across all the corpora. The results in Table 4 show that HAF significantly improved the performance of emotion recognition when compared to the methods of the SAVEE MFCC related features [42] and RAVDESS wavelet transform features [49] that recorded the highest recognition accuracies of 76.40% and 60.10% respectively.

In particular, the results of precision, recall, F1-score and accuracy per each emotion, learning algorithm and database were subsequently discussed. Table 5 shows that angry, calm and disgust emotions were easier to recognize using RDF and GBM ensemble learning of RAVDESS MFCC1 features. The results were particularly impressive with RDF ensemble learning with perfect precision, recall and F1-score computed by the algorithm, which also recorded highest accuracy values across all emotions. However, BMLP generally gave the least performing result across all emotions. RAVDESS MFCC1 features achieved a precision rate of 92.0% for the fear emotion using RDF ensemble learning, which is relatively high when compared to the results in [27].

Table 5. Percentage precision, recall and F1-score and accuracy on RAVDESS MFCC1.

Classifier/Measure	Angry	Calm	Disgust	Fear	Happy	Neutral	Sad	Surprise
			RDF					
Precision	100.0	100.0	100.0	92.0	80.0	53.0	87.0	96.0
Recall	100.0	100.0	100.0	83.0	82.0	86.0	79.0	91.0
F1-score	100.0	100.0	100.0	87.0	81.0	66.0	83.0	94.0
Accuracy	91.0	90.0	91.0	89.0	91.0	89.0	91.0	93.0
			GBM					
Precision	94.0	100.0	100.0	87.0	69.0	39.0	87.0	85.0
Recall	100.0	100.0	100.0	74.0	69.0	78.0	72.0	87.0
F1-score	97.0	100.0	100.0	80.0	69.0	52.0	79.0	86.0
Accuracy	86.0	85.2	87.0	81.0	83.0	85.0	86.0	90.0
			ABC					
Precision	91.0	82.0	78.0	79.0	81.0	79.0	81.0	81.0
Recall	100.0	100.0	100.0	81.0	56.0	85.0	71.0	82.0
F1-score	95.0	100.0	91.0	65.0	69.0	70.0	81.0	79.0
Accuracy	85.0	84.0	85.0	75.0	81.0	79.0	87.0	88.0
			BSVM					
Precision	83.0	84.0	82.0	75.0	75.0	77.0	79.0	89.0
Recall	82.0	83.0	91.0	73.0	79.0	81.0	84.0	87.0
F1-score	81.0	82.0	79.0	79.0	81.0	79.0	82.0	84.0
Accuracy	83.0	85.0	81.0	79.0	80.6	81.0	86.0	86.0
			BMLP					
Precision	72.0	81.0	78.0	68.0	70.0	70.0	75.0	81.0
Recall	78.0	84.0	82.0	70.0	76.0	79.0	78.0	83.0
F1-score	74.0	82.0	80.0	67.0	71.0	75.0	77.0	80.0
Accuracy	80.0	77.0	76.0	71.0	74.0	75.0	75.0	77.0

Table 6 shows the results of percentage precision, recall, F1-score and accuracy performance analysis of RAVDESS MFCC2 features. The performance values were generally higher for recognizing angry, calm and disgust emotions using RDF. However, the performance values were relatively low for happy and neutral emotions showing that the hybridization of MFCC, ZCR, energy and fundamental frequency features yielded a relatively poor result in recognizing happy and neutral emotions using ensemble learning. The MFCC2 combination of acoustic features achieved an improved recognition performance of the fear (90.0%) emotion using RDF in comparison with the report in [44], where 81.0% recognition was achieved for the fear emotion using the CASIA database.

Table 6. Percentage precision, recall and F1-score and accuracy on RAVDESS MFCC2.

Classifier/Measure	Angry	Calm	Disgust	Fear	Happy	Neutral	Sad	Surprise
				Emotion				
			RDF					
Precision	100.0	100.0	100.0	91.0	95.0	64.0	90.0	100.0
Recall	100.0	100.0	100.0	89.0	78.0	100.0	90.0	95.0
F1-score	100.0	100.0	100.0	93.0	86.0	78.0	90.0	98.0
Accuracy	96.0	93.0	92.0	90.0	94.0	91.0	96.0	97.6
			GBM					
Precision	100.0	100.0	100.0	87.0	79.0	55.0	80.0	86.0
Recall	100.0	100.0	100.0	87.0	75.0	86.0	67.0	90.0
F1-score	100.0	100.0	100.0	87.0	77.0	67.0	73.0	88.0
Accuracy	88.0	84.0	83.0	82.0	85.0	85.0	88.0	98.6
			ABC					
Precision	94.0	93.0	91.0	78.0	71.0	79.0	81.0	89.0
Recall	96.0	100.0	100.0	86.0	76.0	73.0	83.0	86.0
F1-score	92.0	95.0	100.0	79.0	74.0	76.0	81.0	86.0
Accuracy	86.0	82.0	83.0	81.0	87.0	82.0	88.0	92.6
			BSVM					
Precision	81.0	94.0	91.0	77.0	74.0	79.0	78.0	94.0
Recall	92.0	89.0	90.0	76.0	74.0	77.0	74.0	90.0
F1-score	94.0	91.0	91.0	77.0	74.0	78.0	76.0	92.0
Accuracy	84.0	81.0	80.0	78.0	85.0	85.0	87.0	96.0
			BMLP					
Precision	76.0	89.0	86.0	72.0	70.0	76.0	74.0	83.0
Recall	75.0	86.0	86.0	74.0	67.0	75.0	72.0	85.0
F1-score	77.0	91.0	87.0	74.0	73.0	76.0	74.0	90.0
Accuracy	85.0	81.0	78.0	73.0	78.0	77.0	78.0	84.0

Emotion recognition results shown in Table 7 with respect to the RAVDESS HAF were generally impressive across emotions, irrespective of the ensemble learning algorithms utilized. RDF was highly effective in recognizing all the eight emotions because it had achieved perfect accuracy value by 100.0% on almost all the emotions, except fear (98.0%). Moreover, perfect precision and accuracy values were achieved in recognizing the neutral emotion, which was really difficult to identify using other sets of features. In addition, 98.0% accuracy and perfect values obtained for precision, recall and F1-score for the fear emotion is impressive because of the difficulty of recognizing the fear emotion as reported in the literature [25–27]. All the other learning algorithms performed relatively well, considering the fact that noise was added to the training data. These results attest to the effectiveness of the proposed HAF for recognizing emotions in human speech. In addition, the results point to the significance of sourcing highly discriminating features for recognizing human emotions. They further show that the fear emotion could be recognized efficiently when compared to other techniques used in [44]. It can be seen that the recognition of fear emotion was successfully improved using the proposed HAF. In addition, HAF gave better performance than the bottleneck features used for six emotions from CASIA database [27] that achieved a recognition rate of 60.5% for the fear emotion.

Table 7. Percentage precision, recall and F1-score and accuracy on RAVDESS HAF.

Classifier/Measure	Angry	Calm	Disgust	Fear	Happy	Neutral	Sad	Surprise
				Emotion				
			RDF					
Precision	100.0	100.0	100.0	100.0	100.0	100.0	100.0	98.0
Recall	100.0	100.0	100.0	100.0	100.0	96.0	100.0	100.0
F1-score	100.0	100.0	100.0	100.0	100.0	98.0	100.0	99.0
Accuracy	100.0	100.0	100.0	98.0	100.0	100.0	100.0	100.0
			GBM					
Precision	93.0	100.0	100.0	94.0	100.0	95.0	99.0	93.0
Recall	92.0	100.0	100.0	93.0	97.0	100.0	76.0	100.0
F1-score	92.0	100.0	100.0	93.0	98.0	97.0	86.0	96.0
Accuracy	95.0	91.0	92.0	88.0	91.0	90.0	96.0	98.0
			ABC					
Precision	92.0	100.0	96.0	94.0	100.0	95.0	93.0	97.0
Recall	91.0	100.0	100.0	95.0	93.0	76.0	100.0	100.0
F1-score	92.0	100.0	98.0	93.0	95.0	91.0	87.0	98.0
Accuracy	93.0	90.0	92.0	86.0	93.0	91.0	95.0	96.0
			BSVM					
Precision	91.0	100.0	99.0	92.0	89.0	74.0	83.0	94.0
Recall	91.0	100.0	100.0	90.0	86.0	85.0	86.0	98.0
F1-score	93.0	100.0	100.0	91.0	86.0	77.0	87.0	95.0
Accuracy	93.0	93.0	92.0	86.0	90.0	91.0	93.0	96.0
			BMLP					
Precision	92.0	100.0	93.0	90.0	88.0	73.0	81.0	91.0
Recall	91.0	100.0	97.0	93.0	88.0	86.0	88.0	97.0
F1-score	90.0	100.0	91.0	88.0	83.0	75.0	84.0	91.0
Accuracy	92.0	91.0	93.0	89.0	91.0	88.0	92.0	94.0

Table 8 shows the results of percentage precision, recall, F1-score and accuracy analysis of SAVEE MFCC1 features. In the results, the maximum percentage precision F1-score of the neutral emotion had achieved 94.0% using the RDF ensemble algorithm. This F1-score was quite high when compared to the F1-score obtained from RAVDESS database using the same features. The percentage increase in overall precision of neutral emotion was observed to be 133.33%, 77.36%, 18.57%, 12.66% and 10.39% using GBM, RDF, BMLP, ABC and BSVM respectively. Moreover, RDF and GBM algorithms achieved perfect prediction (100.0%) for recognizing the surprise emotion. The results show that MFCC features still achieved low recognition rates for the fear emotion as reported by other authors [25]. However, RDF (94.0%) and GBM (91.0%) show higher precision rates for the neutral emotion, which is higher than what was reported in [11].

Table 8. Percentage precision, recall and F1-score and accuracy on SAVEE MFCC1.

Classifier/Measure	Angry	Disgust	Fear	Happy	Neutral	Sad	Surprise
			RDF				
Precision	65.0	24.0	65.0	48.0	94.0	48.0	100.0
Recall	65.0	83.0	57.0	65.0	52.0	83.0	100.0
F1-score	65.0	37.0	60.0	55.0	67.0	61.0	100.0
Accuracy	69.0	66.0	60.0	75.0	64.0	77.0	81.0
			GBM				
Precision	65.0	29.0	75.0	26.0	91.0	62.0	100.0
Recall	62.0	67.0	48.0	55.0	62.0	72.0	100.0
F1-score	63.0	40.0	59.0	35.0	74.0	67.0	100.0
Accuracy	70.0	66.0	57.0	74.0	62.0	79.0	84.0
			ABC				
Precision	66.0	28.0	64.0	36.0	89.0	59.0	98.0
Recall	68.0	29.0	62.0	40.0	91.0	59.0	100.0
F1-score	65.0	28.0	63.0	38.0	91.0	58.0	99.0
Accuracy	60.0	57.0	53.0	66.0	62.0	79.0	83.0
			BSVM				
Precision	65.0	23.0	61.0	41.0	85.0	57.0	98.0
Recall	64.0	26.0	57.0	38.0	81.0	61.0	100.0
F1-score	66.0	29.0	62.0	36.0	83.0	58.0	99.0
Accuracy	56.0	54.0	51.0	69.0	56.0	77.0	85.0
			BMLP				
Precision	64.0	23.0	59.0	38.0	83.0	54.0	95.0
Recall	63.0	25.0	55.0	39.0	85.0	57.0	97.0
F1-score	64.0	25.0	59.0	37.0	81.0	55.0	96.0
Accuracy	52.0	50.0	49.0	51.0	53.0	64.0	69.0

Table 9 shows the analysis results of the percentage precision, recall, F1-score and accuracy obtained after performing the classification with SAVEE MFCC2 features. It can be observed from the table that all the learning algorithms performed well for recognizing the surprise emotion, but performed poorly in recognizing other emotions. In general, all learning algorithms achieved a low prediction of emotions with the exception of surprise emotion in which the perfect F1-score was obtained using the RDF algorithm. The lowest precision value of 93.0% was achieved for the surprise emotion using MFCC2 features with BMLP. These results indicate the unsuitability of MFCC, ZCR, energy and fundamental frequency features for recognizing human emotions. However, RDF and GBM gave higher recognition results for happy and sad emotions, RDF gave a higher recognition result for fear emotion while ABC and BSVM gave higher recognition results for the sad emotion that the traditional and bottleneck features [27].

Table 9. Percentage precision, recall and F1-score and accuracy on SAVEE MFCC2.

Classifier/Measure	Angry	Disgust	Fear	Happy	Neutral	Sad	Surprise
			RDF				
Precision	71.0	85.0	69.0	68.0	95.0	59.0	100.0
Recall	70.0	84.0	65.0	67.0	68.0	88.0	100.0
F1-score	71.0	84.0	68.0	68.0	74.0	71.0	100.0
Accuracy	77.0	73.0	63.0	70.0	69.0	82.0	100.0
			GBM				
Precision	72.0	65.0	83.0	47.0	93.0	69.0	100.0
Recall	69.0	74.0	69.0	62.0	71.0	77.0	100.0
F1-score	70.0	68.0	71.0	48.0	77.0	75.0	100.0
Accuracy	74.0	71.0	59.0	73.0	69.0	86.0	92.0
			ABC				
Precision	74.0	71.0	70.0	58.0	69.0	73.0	99.0
Recall	76.0	72.0	68.0	61.0	68.0	71.0	100.0
F1-score	74.0	75.0	73.0	59.0	65.0	73.0	100.0
Accuracy	71.0	67.0	56.0	69.0	69.0	71.0	99.0
			BSVM				
Precision	71.0	68.0	66.0	59.0	67.0	71.0	98.0
Recall	72.0	72.0	64.0	61.0	64.0	75.0	100.0
F1-score	70.0	70.0	66.0	60.0	65.0	73.0	100.0
Accuracy	71.0	62.0	51.0	67.0	68.0	73.0	100.0
			BMLP				
Precision	70.0	66.0	64.0	57.0	66.0	67.0	93.0
Recall	72.0	70.0	66.0	59.0	68.0	72.0	97.0
F1-score	71.0	68.0	64.0	59.0	65.0	69.0	96.0
Accuracy	56.0	52.0	49.0	65.0	54.0	64.0	85.0

Table 10 shows the analysis results of the percentage precision, recall, F1-score and accuracy obtained after performing classification with SAVEE HAF. Accordingly, all the emotions were much easier to recognize using the HAF that had tremendously improved the recognition of the fear emotion reported in the literature to be difficult with lower performance results recorded with traditional and bottleneck features for fear emotion [27]. Moreover, Kerkeni et al. [11] obtained a recognition accuracy rate of 76.16% for the fear emotion using MFCC and MS features based on the Spanish database of seven emotions. These results show that using HAF with ensemble learning is highly promising for recognizing emotions in human speech. The study findings were consistent with the literature that ensemble learning gives a better predictive performance through the fusion of information knowledge of predictions of multiple inducers [75,85]. The ability of ensemble learning algorithms to mimic the nature of human by seeking opinions from several inducers for informed decision distinguish them from inducers [73]. Moreover, it can be seen across Tables 4–10 that the set of HAF presented the most effective acoustic features for speech emotion recognition, while the set MFCC1 presented the worst features. In addition, the study results supported the hypothesis that a combination of acoustic features based on prosodic and spectral reflects the important characteristics of speech [42,53,54].

Table 10. Percentage precision, recall and F1-score and accuracy on SAVEE HAF.

Classifier/Measure	Angry	Disgust	Fear	Happy	Neutral	Sad	Surprise
			RDF				
Precision	100.0	100.0	94.0	100.0	100.0	100.0	100.0
Recall	100.0	94.0	100.0	100.0	100.0	100.0	100.0
F1-score	100.0	100.0	100.0	100.0	100.0	97.0	97.0
Accuracy	100.0	99.0	97.0	100.0	99.1	100.0	100.0
			GBM				
Precision	100.0	96.0	100.0	100.0	100.0	100.0	100.0
Recall	100.0	100.0	94.0	100.0	100.0	100.0	100.0
F1-score	100.0	98.0	97.0	100.0	100.0	100.0	100.0
Accuracy	100.0	98.1	99.0	100.0	100.0	100.0	100.0
			ABC				
Precision	100.0	94.0	100.0	100.0	100.0	100.0	100.0
Recall	100.0	100.0	96.0	100.0	100.0	100.0	100.0
F1-score	100.0	94.0	91.0	100.0	100.0	100.0	100.0
Accuracy	100.0	99.0	99.0	100.0	100.0	100.0	100.0
			BSVM				
Precision	100.0	93.0	99.0	99.0	100.0	100.0	100.0
Recall	100.0	96.0	98.0	99.0	100.0	100.0	100.0
F1-score	100.0	96.0	99.0	100.0	100.0	100.0	100.0
Accuracy	100.0	94.0	96.0	96.0	100.0	100.0	100.0
			BMLP				
Precision	100.0	90.0	98.0	97.0	100.0	100.0	100.0
Recall	100.0	93.0	97.0	99.0	100.0	100.0	100.0
F1-score	100.0	91.0	98.0	98.0	100.0	100.0	100.0
Accuracy	100.0	93.0	95.0	99.0	100.0	100.0	100.0

Tables 4–10 show that HAF features had a higher discriminating power because higher performance results were achieved across all emotion classes in the SAVEE and RAVDESS corpora. The performance of each ensemble learning algorithm was generally low when MFCC1 and MFCC2 were applied because the sets of the features had a much lower discriminating power compared to HAF. The recognition results across the experimental databases were different because RAVDESS had more instances than SAVEE. Moreover, SAVEE constitutes male speakers only while RAVDESS is comprised of both male and female speakers. According to the literature, this diversity has an impact on the performance of emotion recognition [35]. The literature review revealed that most emotion recognition models have not been able to recognize the fear emotion with a higher classification accuracy of 98%, but the results in Table 4 show that HAF significantly improved the recognition performance of emotion beyond expectation.

5. Conclusions

The automatic recognition of emotion is still an open research because human emotions are highly influenced by quite a number of external factors. The primary contribution of this study is the construction and validation of a set of hybrid acoustic features based on prosodic and spectral features for improving speech emotion recognition. The constructed acoustic features have made it easy to effectively recognize eight classes of human emotions as seen in this paper. The agglutination of prosodic and spectral features has been demonstrated in this study to yield excellent classification results. The proposed set of acoustic features is highly effective in recognizing all the eight emotions investigated in this study, including the fear emotion that has been reported in the literature to be difficult to classify. The proposed methodology seems to work well because combining certain prosodic and spectral features increases the overall discriminating power of the features hence increasing classification accuracy. This superiority is further underlined by the fact that the same ensemble learning algorithms were used on different feature sets, exposing the performance of each group of features regarding their discriminating power. In addition, we saw that ensemble learning algorithms generally performed well. They are widely judged to improve recognition performance better than a single inducer by decreasing variability and reducing bias. Experimental results show that it was quite difficult to achieve high precisions and accuracies when recognizing the neutral emotion with either pure MFCC features or a combination of MFCC, ZCR, energy and fundamental frequency features that were seen to be effective in recognizing the surprise emotion. However, we provided evidence through intensive experimentation that random decision forest ensemble learning of the proposed hybrid acoustic features was highly effective for speech emotion recognition.

The results of this study were generally fantastic, even though the Gaussian noise was added to the training data. The gradient boosting machine algorithm yielded good results, but the main challenge with the algorithm is that its training time is long. In terms of effectiveness, the random decision forest algorithm is superior to the gradient boosting machine algorithm for speech emotion recognition with the proposed acoustic features. The results obtained in this study were quite stimulating, nevertheless, the main limitation of this study was that the experiments were done on acted speech databases in consonance with the research culture. There can be major differences between working with acted and real data. Moreover, there are new features such as breath and spectrogram not investigated in this study for recognizing speech emotion. In the future, we would like to pursue this direction to determine how the new features will compare with the proposed hybrid acoustic features. In addition, future research will harvest real emotion data for emotion recognition using the Huawei OceanConnect cloud based IoT platform and Narrow Band IoT resources in South Africa Luban workshop at the institution of the authors, which is a partnership project with the Tianjin Vocational Institute in China.

Author Contributions: Conceptualization by K.Z. and O.O., Data curation by K.Z. Methodology by K.Z. Writing original draft by K.Z. and O.O. Writing review and editing by O.O., Supervision by O.O. All authors have read and agreed to the published version of the manuscript.

Funding: This research received no external funding.

Conflicts of Interest: The authors declare no conflict of interest.

References

1. Ma, Y.; Hao, Y.; Chen, M.; Chen, J.; Lu, P.; Košir, A. Audio-visual emotion fusion (AVEF): A deep efficient weighted approach. *Inf. Fusion* **2019**, *46*, 184–192. [CrossRef]
2. Picard, R.W. Affective computing: Challenges. *Int. J. Hum. Comput. Stud.* **2003**, *59*, 55–64. [CrossRef]
3. Sarker, K.; Alam, K.R. Emotion recognition from human speech: Emphasizing on relevant feature selection and majority voting technique. In Proceedings of the 3rd International Conference on Informatics, Electronics & Vision (ICIEV), Dhaka, Bangladesh, 23–24 May 2014; pp. 89–95.
4. Li, S.; Xu, L.; Yang, Z. Multidimensional speaker information recognition based on proposed baseline system. In Proceedings of the 2017 IEEE 2nd Advanced Information Technology, Electronic and Automation Control Conference (IAEAC), Chongqing, China, 25–26 March 2017; pp. 1776–1780.
5. Jiang, H.; Hu, B.; Liu, Z.; Yan, L.; Wang, T.; Liu, F.; Kang, H.; Li, X. Investigation of different speech types and emotions for detecting depression using different classifiers. *Speech Commun.* **2017**, *90*, 39–46. [CrossRef]
6. Subhashini, R.; Niveditha, P.R. Analyzing and detecting employee's emotion for amelioration of organizations. *Procedia Comput. Sci.* **2015**, *48*, 530–536. [CrossRef]

7. Feinberg, R.A.; Hokama, L.; Kadam, R.; Kim, I. Operational determinants of caller satisfaction in the banking/financial services call center. *Int. J. Bank Mark.* **2002**, *20*, 174–180. [CrossRef]
8. Gomes, J.; El-Sharkawy, M. Implementation of i-vector algorithm in speech emotion recognition by using two different classifiers: Gaussian mixture model and support vector machine. *Int. J. Adv. Res. Comput. Sci. Softw. Eng.* **2016**, *6*, 8–16.
9. Yu, T.-K.; Fang, K. Measuring the post-adoption customer perception of mobile banking services. *Cyberpsychol. Behav.* **2009**, *12*, 33–35. [CrossRef]
10. Chakraborty, R.; Pandharipande, M.; Kopparapu, S.K. Knowledge-based framework for intelligent emotion recognition in spontaneous speech. *Procedia Comput. Sci.* **2016**, *96*, 587–596. [CrossRef]
11. Kerkeni, L.; Serrestou, Y.; Mbarki, M.; Raoof, K.; Mahjoub, M.A. Speech emotion recognition: Methods and cases study. In Proceedings of the 10th International Conference on Agents and Artificial Intelligence (ICAART 2018), Funchal, Madeira, Portugal, 16–18 January 2018; Volume 2, pp. 175–182.
12. Arias, J.P.; Busso, C.; Yoma, N.B. Shape-based modeling of the fundamental frequency contour for emotion detection in speech. *Comput. Speech Lang.* **2014**, *28*, 278–294. [CrossRef]
13. Atayero, A.A.; Olugbara, O.O.; Ayo, C.K.; Ikhu-Omoregbe, N.A. Design, development and deployment of an automated speech-controlled customer care service system. In Proceedings of the GSPx, The International Embedded Solutions Event, Santa Clara, CA, USA, 27–30 September 2004; pp. 27–30.
14. Hess, U.; Thibault, P. Darwin and emotion expression. *Am. Psychol.* **2009**, *64*, 120–128. [CrossRef]
15. Ying, S.; Xue-Ying, Z. Characteristics of human auditory model based on compensation of glottal features in speech emotion recognition. *Future Gener. Comput. Syst.* **2018**, *81*, 291–296. [CrossRef]
16. Xiaoqing, J.; Kewen, X.; Yongliang, L.; Jianchuan, B. Noisy speech emotion recognition using sample reconstruction and multiple-kernel learning. *J. China Univ. Posts Telecommun.* **2017**, *24*, 17–19. [CrossRef]
17. Papakostas, M.; Spyrou, E.; Giannakopoulos, T.; Siantikos, G.; Sgouropoulos, D.; Mylonas, P.; Makedon, F. Deep visual attributes vs. hand-crafted audio features on multidomain speech emotion recognition. *Computation* **2017**, *5*, 26. [CrossRef]
18. Arruti, A.; Cearreta, I.; Álvarez, A.; Lazkano, E.; Sierra, B. Feature selection for speech emotion recognition in Spanish and Basque: On the use of machine learning to improve human-computer interaction. *PLoS ONE* **2014**, *9*, e108975. [CrossRef]
19. Altun, H.; Polat, G. Boosting selection of speech related features to improve performance of multi-class SVMs in emotion detection. *Expert Syst. Appl.* **2009**, *36*, 8197–8203. [CrossRef]
20. Luengo, I.; Navas, E.; Hernaez, I. Feature analysis and evaluation for automatic emotion identification in speech. *IEEE Trans. Multimed.* **2010**, *12*, 490–501. [CrossRef]
21. Basu, S.; Chakraborty, J.; Bag, A.; Aftabuddin, M. A review on emotion recognition using speech. In Proceedings of the 2017 International Conference on Inventive Communication and Computational Technologies (ICICCT), Coimbatore, India, 10–11 March 2017; pp. 109–114.
22. Cong, P.; Wang, C.; Ren, Z.; Wang, H.; Wang, Y.; Feng, J. Unsatisfied customer call detection with deep learning. In Proceedings of the 2016 10th International Symposium on Chinese Spoken Language Processing (ISCSLP), Tianjin, China, 17–20 October 2016; pp. 1–5.
23. Getahun, F.; Kebede, M. Emotion identification from spontaneous communication. In Proceedings of the 2016 12th International Conference on Signal-Image Technology & Internet-Based Systems (SITIS), Naples, Italy, 28 November–1 December 2016; pp. 151–158.
24. Palo, H.K.; Chandra, M.; Mohanty, M.N. Emotion recognition using MLP and GMM for Oriya language. *Int. J. Comput. Vis. Robot.* **2017**, *7*, 426–442. [CrossRef]
25. Khan, A.; Roy, U.K. Emotion recognition using prosodie and spectral features of speech and Naïve Bayes classifier. In Proceedings of the 2017 International Conference on Wireless Communications, Signal Processing and Networking (WiSPNET), Chennai, India, 22–24 March 2017; pp. 1017–1021.
26. Semwal, N.; Kumar, A.; Narayanan, S. Automatic speech emotion detection system using multi-domain acoustic feature selection and classification models. In Proceedings of the 2017 IEEE International Conference on Identity, Security and Behavior Analysis (ISBA), New Delhi, India, 22–24 February 2017; pp. 1–6.
27. Sun, L.; Zou, B.; Fu, S.; Chen, J.; Wang, F. Speech emotion recognition based on DNN-decision tree SVM model. *Speech Commun.* **2019**, *115*, 29–37. [CrossRef]
28. Bhaskar, J.; Sruthi, K.; Nedungadi, P. Hybrid approach for emotion classification of audio conversation based on text and speech mining. *Procedia Comput. Sci.* **2015**, *46*, 635–643. [CrossRef]

29. Liu, Z.T.; Wu, M.; Cao, W.H.; Mao, J.W.; Xu, J.P.; Tan, G.Z. Speech emotion recognition based on feature selection and extreme learning machine decision tree. *Neurocomputing* **2018**, *273*, 271–280. [CrossRef]
30. Cao, H.; Verma, R.; Nenkova, A. Speaker-sensitive emotion recognition via ranking: Studies on acted and spontaneous speech. *Comput. Speech Lang.* **2015**, *29*, 186–202. [CrossRef] [PubMed]
31. Stuhlsatz, A.; Meyer, C.; Eyben, F.; Zielke, T.; Meier, G.; Schuller, B. Deep neural networks for acoustic emotion recognition: Raising the benchmarks. In Proceedings of the 2011 IEEE International Conference on Acoustics, Speech and Signal Processing (ICASSP), Prague, Czech Republic, 22–27 May 2011; pp. 5688–5691.
32. Zhang, Z.; Coutinho, E.; Deng, J.; Schuller, B. Cooperative learning and its application to emotion recognition from speech. *IEEE/ACM Trans. Audio Speech Lang. Process.* **2015**, *23*, 115–126. [CrossRef]
33. Sun, Y.; Wen, G.; Wang, J. Weighted spectral features based on local Hu moments for speech emotion recognition. *Biomed. Signal Process. Control* **2015**, *18*, 80–90. [CrossRef]
34. Shaqra, F.A.; Duwairi, R.; Al-Ayyoub, M. Recognizing emotion from speech based on age and gender using hierarchical models. *Procedia Comput. Sci.* **2019**, *151*, 37–44. [CrossRef]
35. Pérez-Espinosa, H.; Reyes-García, C.A.; Villaseñor-Pineda, L. Acoustic feature selection and classification of emotions in speech using a 3D continuous emotion model. *Biomed. Signal Process. Control* **2012**, *7*, 79–87. [CrossRef]
36. Jiang, W.; Wang, Z.; Jin, J.S.; Han, X.; Li, C. Speech emotion recognition with heterogeneous feature unification of deep neural network. *Sensors* **2019**, *19*, 2730. [CrossRef]
37. Prasada Rao, K.; Chandra Sekhara Rao, M.V.P.; Hemanth Chowdary, N. An integrated approach to emotion recognition and gender classification. *J. Vis. Commun. Image Represent.* **2019**, *60*, 339–345. [CrossRef]
38. Narendra, N.P.; Alku, P. Dysarthric speech classification from coded telephone speech using glottal features. *Speech Commun.* **2019**, *110*, 47–55. [CrossRef]
39. Alonso, J.B.; Cabrera, J.; Medina, M.; Travieso, C.M. New approach in quantification of emotional intensity from the speech signal: Emotional temperature. *Expert Syst. Appl.* **2015**, *42*, 9554–9564. [CrossRef]
40. Alshamsi, H.; Kepuska, V.; Alshamsi, H.; Meng, H. Automated facial expression and speech emotion recognition app development on smart phones using cloud computing. In Proceedings of the 2018 IEEE 9th Annual Information Technology, Electronics and Mobile Communication Conference (IEMCON), Vancouver, BC, Canada, 1–3 November 2018; pp. 730–738.
41. Li, X.; Akagi, M. Improving multilingual speech emotion recognition by combining acoustic features in a three-layer model. *Speech Commun.* **2019**, *110*, 1–12. [CrossRef]
42. Liu, Z.T.; Xie, Q.; Wu, M.; Cao, W.H.; Mei, Y.; Mao, J.W. Speech emotion recognition based on an improved brain emotion learning model. *Neurocomputing* **2018**, *309*, 145–156. [CrossRef]
43. Mao, Q.; Xu, G.; Xue, W.; Gou, J.; Zhan, Y. Learning emotion-discriminative and domain-invariant features for domain adaptation in speech emotion recognition. *Speech Commun.* **2017**, *93*, 1–10. [CrossRef]
44. Wang, K.; An, N.; Li, B.N.; Zhang, Y.; Li, L. Speech emotion recognition using Fourier parameters. *IEEE Trans. Affect. Comput.* **2015**, *6*, 69–75. [CrossRef]
45. Muthusamy, H.; Polat, K.; Yaacob, S. Improved emotion recognition using gaussian mixture model and extreme learning machine in speech and glottal signals. *Math. Probl. Eng.* **2015**, *2015*, 13. [CrossRef]
46. Zhu, L.; Chen, L.; Zhao, D.; Zhou, J.; Zhang, W. Emotion recognition from chinese speech for smart affective services using a combination of SVM and DBN. *Sensors* **2017**, *17*, 1694. [CrossRef]
47. Álvarez, A.; Sierra, B.; Arruti, A.; Lópezgil, J.M.; Garay-Vitoria, N. Classifier subset selection for the stacked generalization method applied to emotion recognition in speech. *Sensors* **2016**, *16*, 21. [CrossRef]
48. Bhavan, A.; Chauhan, P.; Shah, R.R. Bagged support vector machines for emotion recognition from speech. *Knowl. Based Syst.* **2019**, *184*, 104886. [CrossRef]
49. Shegokar, P.; Sircar, P. Continuous wavelet transform based speech emotion recognition. In Proceedings of the 2016 10th International Conference on Signal Processing and Communication Systems (ICSPCS), Gold Coast, QLD, Australia, 19–21 December 2016; pp. 1–8.
50. Kerkeni, L.; Serrestou, Y.; Raoof, K.; Mbarki, M.; Mahjoub, M.A.; Cleder, C. Automatic speech emotion recognition using an optimal combination of features based on EMD-TKEO. *Speech Commun.* **2019**, *114*, 22–35. [CrossRef]
51. Livingstone, S.R.; Russo, F.A. The ryerson audio-visual database of emotional speech and song (ravdess): A dynamic, multimodal set of facial and vocal expressions in north American english. *PLoS ONE* **2018**, *13*, 1–38. [CrossRef]

52. Wang, W.; Klinger, K.; Conapitski, C.; Gundrum, T.; Snavely, J. *Machine Audition: Principles, Algorithms, and Systems*; IGI Global Press: Hershey, PA, USA, 2010.
53. Ibrahim, N.J.; Idris, M.Y.I.; Yakub, M.; Yusoff, Z.M.; Rahman, N.N.A.; Dien, M.I. Robust feature extraction based on spectral and prosodic features for classical Arabic accents recognition. *Malaysian J.Comput. Sci.* **2019**, 46–72.
54. Banse, R.; Scherer, K.R. Acoustic profiles in vocal emotion expression. *J. Pers. Soc. Psychol.* **1996**, *70*, 614–636. [CrossRef] [PubMed]
55. McEnnis, D.; McKay, C.; Fujinaga, I.; Depalle, P. jAudio: A feature extraction library. In Proceedings of the International Conference on Music Information Retrieval, London, UK, 11–15 September 2005; pp. 600–603.
56. Hellbernd, N.; Sammler, D. Prosody conveys speaker's intentions: Acoustic cues for speech act perception. *J. Mem. Lang.* **2016**, *88*, 70–86. [CrossRef]
57. El Ayadi, M.; Kamel, M.S.; Karray, F. Survey on speech emotion recognition: Features, classification schemes, and databases. *Pattern Recognit.* **2011**, *44*, 572–587. [CrossRef]
58. Guidi, A.; Gentili, C.; Scilingo, E.P.; Vanello, N. Analysis of speech features and personality traits. *Biomed. Signal Process. Control.* **2019**, *51*, 1–7. [CrossRef]
59. Pervaiz, M.; Ahmed, T. Emotion recognition from speech using prosodic and linguistic features. *Int. J. Adv. Comput. Sci. Appl.* **2016**, *7*, 84–90. [CrossRef]
60. Chen, L.; Mao, X.; Xue, Y.; Cheng, L.L. Speech emotion recognition: Features and classification models. *Digit. Signal Process.* **2012**, *22*, 1154–1160. [CrossRef]
61. Agostini, G.; Longari, M.; Pollastri, E. Musical instrument timbres classification with spectral features. *EURASIP J. Appl. Signal Process.* **2003**, *2003*, 5–14. [CrossRef]
62. Avisado, H.G.; Cocjin, J.V.; Gaverza, J.A.; Cabredo, R.; Cu, J.; Suarez, M. Analysis of music timbre features for the construction of user-specific affect model. *Theory Pract. Comput.* **2012**, *5*, 28–35.
63. Siedenburg, K.; Fujinaga, I.; McAdams, S. A Comparison of approaches to timbre descriptors in music information retrieval and music psychology. *J. New Music Res.* **2016**, *45*, 27–41. [CrossRef]
64. Istening, E.M.L.; Imbre, T. Embodied listening and timbre: Perceptual, acoustical and neural correlates. *Music Percept.* **2018**, *35*, 332–363.
65. Kos, M.; Kačič, Z.; Vlaj, D. Acoustic classification and segmentation using modified spectral roll-off and variance-based features. *Digit. Signal Process. Rev. J.* **2013**, *23*, 659–674. [CrossRef]
66. Burger, B.; Ahokas, R.; Keipi, A.; Toiviainen, P. Relationships between spectral flux, perceived rhythmic strength, and the propensity to move. In Proceedings of the Sound and Music Computing Conference 2013, SMC 2013, Stockholm, Sweden, 30 July–3 August 2013; pp. 179–184.
67. Rouillard, J.M.; Simler, M. Signal estimation from modified short-time Fourier transform. *Trans. Acoust. Speech Signal Process.* **1952**, *3*, 772–776.
68. Razuri, J.D.; Sundgren, G.; Rahmani, R.; Larsson, A.; Moran, A.; Bonet, I. Speech emotion recognition in emotional feedback for Human-Robot Interaction. *Int. J. Adv. Res. Artif. Intell.* **2015**, *4*, 20–27.
69. Ernawan, F.; Abu, N.A.; Suryana, N. Spectrum analysis of speech recognition via discrete Tchebichef transform. In Proceedings of the International Conference on Graphic and Image Processing (ICGIP 2011), Cairo, Egypt, 1–3 October 2011; Volume 8285, p. 82856L.
70. James, A.P. Heart rate monitoring using human speech spectral features. *Hum. Cent. Comput. Inf. Sci.* **2015**, *5*, 1–12. [CrossRef]
71. Kajarekar, S.; Malayath, N.; Hermansky, H. Analysis of sources of variability in speech. In Proceedings of the Sixth European Conference on Speech Communication and Technology, Budapest, Hungary, 5–9 September 1999.
72. Pachet, F.; Roy, P. Analytical features: A knowledge-based approach to audio feature generation. *EURASIP J. Audio Speech Music. Process.* **2009**, *2009*, 153017. [CrossRef]
73. Turgut, Ö. The acoustic cues of fear: Investigation of acoustic parameters of speech containing fear. *Arch. Acoust.* **2018**, *43*, 245–251.
74. Thakur, S.; Adetiba, E.; Olusgbara, O.O.; Millham, R. Experimentation using short-term spectral features for secure mobile internet voting authentication. *Math. Probl. Eng.* **2015**, *2015*, 564904. [CrossRef]
75. Sagi, O.; Rokach, L. Ensemble learning: A survey. *Wiley Interdiscip. Rev. Data Min. Knowl. Discov.* **2018**, *8*, e1249. [CrossRef]
76. Kotsiantis, S.B.; Pintelas, P.E. Combining bagging and boosting. *Int. J. Comput. Intell.* **2004**, *1*(4), 324–333.

77. de Almeida, R.; Goh, Y.M.; Monfared, R.; Steiner, M.T.A.; West, A. An ensemble based on neural networks with random weights for online data stream regression. *Soft Comput.* **2019**, 1–21. [CrossRef]
78. Huang, M.W.; Chen, C.W.; Lin, W.C.; Ke, S.W.; Tsai, C.F. SVM and SVM ensembles in breast cancer prediction. *PLoS ONE* **2017**, *12*, e0161501. [CrossRef] [PubMed]
79. Xing, H.J.; Liu, W.T. Robust AdaBoost based ensemble of one-class support vector machines. *Inf. Fusion* **2020**, *55*, 45–58. [CrossRef]
80. Navarro, C.F.; Perez, C. A Color–texture pattern classification using global–local feature extraction, an SVM classifier with bagging ensemble post-processing. *Appl. Sci.* **2019**, *9*, 3130. [CrossRef]
81. Wu, Y.; Ke, Y.; Chen, Z.; Liang, S.; Zhao, H.; Hong, H. Application of alternating decision tree with AdaBoost and bagging ensembles for landslide susceptibility mapping. *Catena* **2020**, *187*, 104396. [CrossRef]
82. Zvarevashe, K.; Olugbara, O.O. Gender voice recognition using random forest recursive feature elimination with gradient boosting machines. In Proceedings of the 2018 International Conference on Advances in Big Data, Computing and Data Communication Systems (icABCD), Durban, South Africa, 6–7 August 2018; pp. 1–6.
83. Yaman, E.; Subasi, A. Comparison of bagging and boosting ensemble machine learning methods for automated EMG signal classification. *BioMed Res. Int.* **2019**, *2019*, 9152506. [CrossRef]
84. Friedman, J. Greedy function approximation: A gradient boosting machine. *Ann. Stat.* **2001**, *29*, 1189. [CrossRef]
85. Dong, X.; Yu, Z.; Cao, W.; Shi, Y.; Ma, Q. A survey on ensemble learning. *Front. Comput. Sci.* **2020**, *14*, 241–258. [CrossRef]
86. Olugbara, O.O.; Taiwo, T.B.; Heukelman, D. Segmentation of melanoma skin lesion using perceptual color difference saliency with morphological analysis. *Math. Probl. Eng.* **2018**, *2018*, 1524286. [CrossRef]
87. Livieris, I.E.; Kotsilieris, T.; Tampakas, V.; Pintelas, P. Improving the evaluation process of students' performance utilizing a decision support software. *Neural Comput. Appl.* **2019**, *31*, 1683–1694. [CrossRef]

© 2020 by the authors. Licensee MDPI, Basel, Switzerland. This article is an open access article distributed under the terms and conditions of the Creative Commons Attribution (CC BY) license (http://creativecommons.org/licenses/by/4.0/).

Article

Ensemble Deep Learning for Multilabel Binary Classification of User-Generated Content

Giannis Haralabopoulos [1,*], **Ioannis Anagnostopoulos** [2] **and Derek McAuley** [1]

1. School of Computer Science, University of Nottingham, Nottingham NG8 1BB, UK; derek.mcauley@nottingham.ac.uk
2. Department of Computer Science and Biomedical Informatics, University of Thessaly, 35131 Lamia, Greece; janag@dib.uth.gr
* Correspondence: giannis.haralabopoulos@nottingham.ac.uk

Received: 14 February 2020; Accepted: 30 March 2020; Published: 1 April 2020

Abstract: Sentiment analysis usually refers to the analysis of human-generated content via a polarity filter. Affective computing deals with the exact emotions conveyed through information. Emotional information most frequently cannot be accurately described by a single emotion class. Multilabel classifiers can categorize human-generated content in multiple emotional classes. Ensemble learning can improve the statistical, computational and representation aspects of such classifiers. We present a baseline stacked ensemble and propose a weighted ensemble. Our proposed weighted ensemble can use multiple classifiers to improve classification results without hyperparameter tuning or data overfitting. We evaluate our ensemble models with two datasets. The first dataset is from Semeval2018-Task 1 and contains almost 7000 Tweets, labeled with 11 sentiment classes. The second dataset is the Toxic Comment Dataset with more than 150,000 comments, labeled with six different levels of abuse or harassment. Our results suggest that ensemble learning improves classification results by 1.5% to 5.4%.

Keywords: ensemble learning; sentiment analysis; multilabel classification; deep neural networks; pure emotion; Semeval 2018 Task 1; toxic comment classification

1. Introduction

Sentiment analysis is the process by which we uncover sentiment from information. The sentiment part could refer to polarity [?], fine grained or not [?], or to pure emotion information [? ? ?]. The most common source of information for sentiment analysis is Online Social Networks (OSNs) [? ?]. User-generated content provides a unique combination of complexity and challenge for automated sentiment classification.

Automated classification refers to methods that can identify and classify information based on an inference process. Machine Learning (ML) studies these types of methods and can be generally separated in three parts: Modeling, Learning, and Classification. Given a classification task, a ML method has to create a model of the data, learn based on a set of pre-classified examples and perform a classification, as required by the task. Ensemble learning refers to the combination of finite number of ML systems to improve the classification results [?].

Various ML systems exist, most frequently characterized by the model and the training methods they employ. Artificial Neural Networks (ANNs) are one type of ML systems [?]. These networks have three layers: input, hidden, and output. In the input layer, data is initially fed into a model. The model parameters are then (re)calculated in the hidden layer, and data is classified in the output layer. Each layer consists of a set of nodes or artificial neurons which are connected to the next layer. When an ANN consists of multiple hidden layers, it is referred to as a Deep Neural Network (DNN) [?].

DNNs have been widely used in computer vision problems [? ? ?], where the goal of the classification is to identify or detect objects/items/features in an image. When the goal of the classification is to detect multiple objects in an image then the task is considered multilabel. These types of problems can be extended from computer vision to text analysis. In emotion related classification, a textual input can convey one or multiple emotions.

Traditional sentiment analysis is focused on a confined polarity or single emotion basis. Our main goal is to present the effectiveness of ensemble learning in text-based multilabel classification. In addition, we aim to trigger the researcher interest for considering multilabel emotion classification as a significant aspect regarding sentiment analysis.

Our contributions are as follows. We create and present five multilabel classification architectures and two ensembles, as well as a baseline stacked ensemble and a weighted ensemble that assigns weights based on differential evolution. Then, we highlight the effectiveness of ensemble learning in modern multilabel emotion datasets. Our results show that ensemble learning can be more effective than single DNN networks in multilabel emotion classification. In addition, we also incorporate a high-level description of the most commonly used hidden layers to introduce readers to deep-learning architectures.

The remainder of our work is formatted as described. Section ?? covers some introductory bibliography alongside state-of-the-art ensemble publications. Section ?? presents in detail our diverse DNN architectures and their individual components. Section ?? describes the ensemble methods we employed as well as some key sub-components. Section ?? presents the datasets we used and some of their properties. Section ?? details our results and potential improvements. Section ?? concludes our study with the summary and future work direction.

2. Related Work

Sentiment analysis is extensively studied since the early 2000s [? ?]. With the advent of internet, OSNs soon became the most used source for sentiment analysis [? ?]. Some of the applications of sentiment analysis are: marketing [?], politics [?], and more recently medicine [? ?]. Affective computing, as suggested by Picard [?], has sparked the interest in the specific emotion analysis of texts [? ?].

Machine learning has been successfully applied to sentiment analysis of texts [? ? ?]. ML methods that have been used for sentiment analysis are: Support Vector Machines [? ?], Multinomial Naïve Bayes [?] and Decision Trees [?], while DNNs were introduced more recently [? ?].

DNNs can be used in text related applications as well. In [?] Severyn and Moschitti rerank short texts in pairs and present the per pair relation without manual feature engineering. Lai et al. [?] perform unsupervised text classification and highlight key text components in the process via a recurrent convolutional neural network. The authors of [?] perform a named entity recognition task and generate relevant word embedding with two separate DNNs. Text generation is addressed via a recurrent neural network in [?] and an extensively trained model outperforms the best non-NN models. Sentiment classification of textual sources had its own fair share of DNN implementations [? ? ?].

Learning ensembles have been used to combine different types of information, such as audio video and text towards sentiment and emotional classification [?]. Araque et al. use an ensemble of classifiers that combines features and word vectors [?] for sentiment classification with greater than 80% F-Score. A soft voting ensemble is used in [?] for topic-document and document classification the results suggesting a significant improvement over single-model methods. The authors of [?] use a stacked two-layer ensemble of CNN to predict the message level sentiment of Tweets, with the addition of a distant supervision phase. A pseudo-ensemble, essentially an ensemble of similar models trained on noisy sub-data, is used for sentiment analysis purposes in [?], but is ineffective for regression classification problems.

Multilabel classification problems assign multiple classes per item. Such problems are frequently observed in the field of computer vision [? ? ?]. With regards to multilabel text-based sentiment analysis, Chen et al. [?] propose an ensemble of a convolution neural network and a recurrent neural network for feature extraction and class prediction correspondingly. The authors of [?] propose a Maximum Entropy model for multilabel classification of short texts found in OSNs. Furthermore, they present an emotion per term lexicon as generated by the model, based on six basic emotions. However, they calculate a micro averaged F1 based on the top emotions per item, essentially converting each weighted label to binary format. Johnson and Zhang [?] present a combination of word order and bag of words in a CNN architecture and point out the threshold sensitivity in multilabel classification.

3. DNN Architectures

We create five different DNNs, with diverse architectures, suited to a multilabel classification problem. Model 1, Figure ??, is a simple CNN with one fully connected layer. Model 2, Figure ??, combines a Gated Recurrent Unit and a Convolution layer, similar to [?]. Model 3, Figure ??, uses Term Frequency Inverse Document Frequency Embeddings and three fully connected layers, inspired by [?]. Model 4, Figure ??, architecture is based on the top performing single model of the Toxic Comment Classification Challenge in Kaggle https://www.kaggle.com/c/jigsaw-toxic-comment-classification-challenge. Model 5, Figure ??, combines uni/bi/tri grams follow by three interconnected CNN processes, as presented in [?]. Each of the modules used in these models is presented in this section.

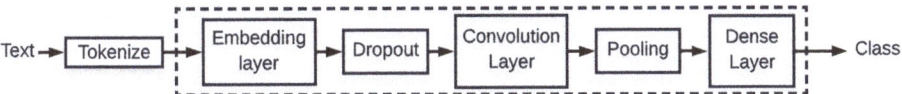

Figure 1. A Convolutional Neural Network.

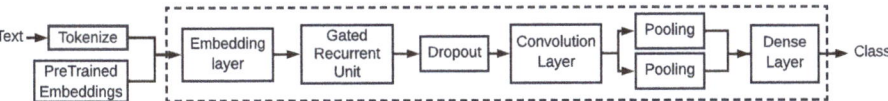

Figure 2. A Recurrent CNN with pre-trained Embeddings.

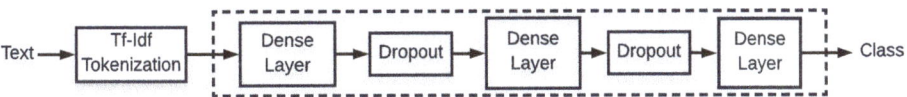

Figure 3. A Deep Neural Network with Tf-Idf Embeddings.

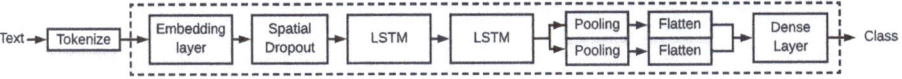

Figure 4. A Long Short-Term Memory network.

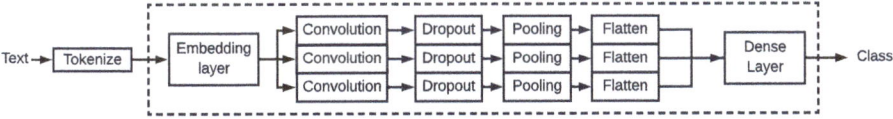

Figure 5. A CNN with Uni/Bi/Trigrams consideration.

3.1. Pre-Processing

For each dataset used, we perform a pre-processing that includes term lemmatization and stemming, lowercase conversion, removal of non-pure text elements (such as Uniform Resource Locators or Emotes), stop word filtering and frequency-based term exclusion. Although some information is lost, extensive term filtering is shown to improve classification results [?].

3.2. Tokenization

Tokenization is performed on a term level for all the five methods presented. Each term is represented by a unique token. Further linguistic elements such as abbreviations and negations are cleaned, returned to their canonical form, and get assigned a token. For example, commonly used negation *'don't'* is tokenized to *'do'*, *'not'*.

3.3. Embedding

We assign a vector value for each token in a sentence, e.g., based on the order they appeared in our corpus, and we create a vector of numerical values. The mapping of word tokens to numerical values of vectors is referred to as embedding. There are various ways of creating word embeddings. Term Frequency-Inverse Document Frequency Tokenization creates a matrix of TF-IDF features which are used to create the embedding. Every sentence is converted to a single dimension vector of numerical elements regardless of the tokenization method. To address variable sentence length, we define a large vector length and we fill the numerical vector with zeroes, a process known as padding. Most common and effective word embedding methods are created based on term co-occurrence throughout a large corpus [? ?].

3.4. Dropout

Neural networks are prone to overfitting. Overfitting is essentially the exhaustive training of the model in a certain set of data, so much that the model fails to generalize. As a result, the model cannot effectively work with new unknown data. A solution to overfitting is to train multiple models and combine their output afterwards, which is highly inefficient.

Srivastava et al. [?] proposed randomly dropping neural units from the network during the training phase. Their results suggested an improvement of the regularization on diverse datasets. Spatial dropout refers to the exact same process, performed over a single axis of elements, rather than random neural units over each layer. Furthermore, dropout has a significant potential to reduce overfitting and provide improvements over other regularization strategies such as L-regularization and soft-weight sharing [?].

3.5. LSTM and Gated Recurrent Unit

A feed-forward neural network has a unidirectional processing path, from input to hidden layer to output. A recurrent network can have information travelling both directions by using feedback loops. Computations derived from earlier input are fed back into the network, imitating human memory. In essence, a recurrent neural network is a chain of identical neural networks that transfer the derived knowledge from one to another. That chain creates learning dependencies that decay mainly due to the size of the chained network.

Hochreiter and Schmidhuber [?] proposed Long Short-Term Memory networks to counter that decay. The novel unit of the LSTM architecture is the memory cell that forgets or remembers the information passed from the previous chain link. The Gated Recurrent Unit of Model 2 was introduced by Cho et al. [?]. Its architecture is similar to LSTM units, but with the absence of an output gate. It is shown that GRU networks perform well on Natural Language Processing tasks [?].

3.6. Convolution

The convolution layer receives as input a tensor, which is convolved and its output, the feature map, is passed to the next layer. A tensor is a mathematical object that describes a mapping of an input set of objects to an output set. Therefore The convolution layers in all models are one-dimensional, with a convolution window of size 3 and an output space dimensionality of 128. The only exception is Model 5, where the convolution windows is different for each layer to provide a differentiated architecture. The primary focus of this layer is to extract features from the input data by preserving the spatial relationship between terms.

3.7. Pooling and Flattening

A pooling layer allows us to reduce the dimensionality of the processed data by keeping only the most important information. Common types of pooling are max, average and sum. The respective operation, per pooling type, is performed in a predefined window of the feature map. A pooling layer reduces the computational cost of the network, leads to scale invariant representations and counters small feature map transformations. Finally, by having a reduced number of features, we decrease the probability of overfitting. Various pooling types were used, based on the architecture. Model 1 uses a single global max pooling layer; the global parameter is outputting the single most important feature of the feature map instead of a feature window. Model 2 uses a combination of global max and a global average pooling layers, while Model 4 uses the local pooling equivalents. Model 5 uses three local max pooling layers, one for each convolution.

The flattening layer on the other hand reduces the dimensionality of the input to one. For example, a feature map with dimensions 5×4 when flattened would produce a one-dimensional vector with 20 elements. The flattening layer passes the most important features of the input data to a fully connected dense layer comprised of the classification neurons.

3.8. Dense

The dense layer is comprised by fully connected neurons both forward and backward. Every element of the input is connected with every neuron of this layer. In four out of five models, a dense layer can be seen at the end of the pipeline. The number of neurons in these layers is the number of classes in our dataset. For the third model, where TF-IDF tokenization takes place, we chose a simple DNN with 3 fully connected layer, which decreasing number of neurons for each subsequent layer. DNNs with multiple dense fully connected layers is shown to perform better than shallow DNNs [?].

3.9. Classification

The output of our model is a multilabel classification vector. Each of the neurons in the final dense layer of the models interact with the classes of the dataset and provide a decimal value (ranging from 0 to 1) which is then rounded for each class. The number of classes defines the number of fully connected neurons in the final dense layer.

4. Ensembles

We previously mentioned that a method to counter overfitting is to train multiple models and then combine their outputs. Ensemble learning combines the single-model outputs to improve predictions and generalization. Ensemble learning improves upon three key aspects of learning, statistics, computation and representation [?]. From a statistics perspective, ensemble methods reduce the risk of data miss-representation, by combining multiple models we reduce the risk of employing a single model trained with biased data. While most learning algorithms search locally for solutions which in turn confines the optimal solution, ensemble methods can execute random seed searches with variable start points with less computational resources. A single hypothesis rarely represents the target function, but an aggregation of multiple hypothesis, as found in ensembles, can better approximate the target function.

We present two ensemble architectures, stacked and weighted [?]. Other popular ensemble methods include AdaBoost, Random Forest and Bagging [?]. Stacked ensembles are the simplest yet one of the most effective ensemble methods, widely used in a variety of applications [? ?]. Stacked ensemble acts as our baseline ensemble, compared with our proposed weighted ensemble based on differential evolution, a meta-heuristic weight optimization method. Meta-heuristic weighted ensembles have achieved remarkable results in single label text classification [? ?].

4.1. Stacked

The main idea behind stacked ensembles is to combine a set of trained models through training of another model (meta-model). The output predictions of the meta-model are based on the training of the model outputs, Algorithm ??. In our implementation we fit the models output into a DNN with two hidden dense layers, Figure ??.

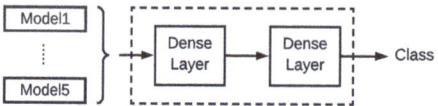

Figure 6. The Stacked Ensemble Architecture.

Algorithm 1 Stacked Ensemble Pseudo-code

1: Input ← Fixed Length Input
2: Output ← Model[1,2,3,4,5] Outputs
3: SE() ← Stacked Ensemble
4: Input → SE(Output) → Class
5: **return** Class

The outputs of the models are merged with the concatenation function of Keras https://keras.io/. The input of the concatenation is a fixed size output tensor of each model. The output of the concatenation is a single tensor, which is then used as an input to the fully connected layer. A second fully connected layer follows similar to the final dense layer on each model.

4.2. Weighted

The weighted ensemble has a similar philosophy behind it. Instead of equally merging the outputs, we merge the outputs by co-calculating a weight. Given a set of weighted tensors and a vector-like object, we sum the product of tensor elements over a single axis, as specified by the one-dimensional vector, Figure ??.

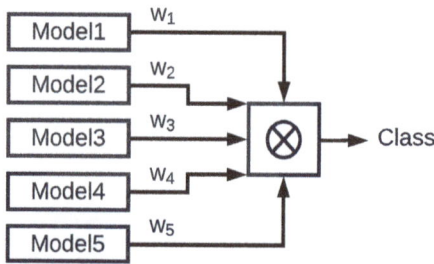

Figure 7. The Weighted Ensemble Architecture.

The 'fitting' process of the second ensemble is a heuristic process for the best possible weights combination, i.e., looking for the global minimum of a multivariate function. We propose differential evolution [?] to scan the large space of five distinct weights, Algorithm ??.

Algorithm 2 Differential Evolution Pseudo-code

1: MF() ← Multivariate Function
2: Target ← Initialize Target Vector
3: Base = Target Vector+Random Integer
4: **for** number of iterations **do**
5:     TempVector1 = Target+RandomDeviation
6:     TempVector2 = Target−RandomDeviation
7:     Weighted = WeightedDifference(TempVector1,TempVector2)
8:     Base = Base+Weighted
9:     Trial = Aggregate(Target,Base)
10:    **if** MF(Trial) < MF(Base) **then**
11:        Base = Trial
12:    **else**
13:        Base = Target
14:    **end if**
15: **end for**
16: **return** Base
---

5. Datasets

For our comparison, we require datasets with binary multilabel categorization. We identified two modern datasets, Semeval 2018 Task 1 Dataset (SEM2018) https://competitions.codalab.org/competitions/17751 and Toxic Comments Dataset from Kaggle (TOXIC) https://www.kaggle.com/c/jigsaw-toxic-comment-classification-challenge.

Both datasets exhibit a level of class imbalance, Figures ??a and ??a. However, they are different not only in context, where SEM2018 is based on Twitter and TOXIC in Wikipedia, but also in the properties of the actual text. The sentence length, after the source is cleaned, is different from the original mainly due to the removal of infrequent terms, Table ??. We discussed before that the dimensions of our term embeddings need to be low. We reduced the dimension by removing the terms that appear no more than 10 times, alongside a tailored stop term removal.

Table 1. Sentence length.

		Original		Cleaned	
		Mean	Median	Mean	Median
SEM2018	Train	16.06	17.0	8.52	9
	Dev.	15.85	16	8.55	9
TOXIC	Train	25.11	15	17.79	11
	Dev.	16.06	17.0	10.22	6

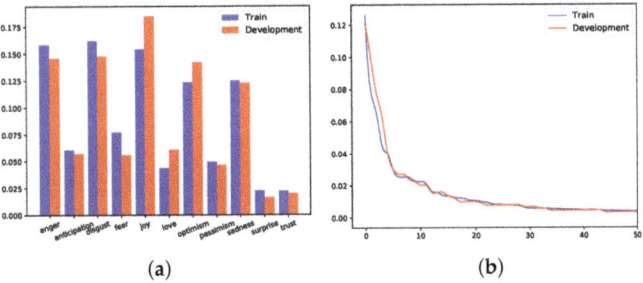

Figure 8. SEM20118 Class Distribution (a) and frequency of unique class combinations (b).

5.1. Semeval 2018

The SEM2018 Train is a collection of 6838 Tweets with emotion labeling of 11 classes. The classes are: *anger, anticipation, disgust, fear, joy, love, optimism, pessimism, sadness, surprise* and *trust*. Some examples of Tweets included in SEM2018 are:

- Whatever you decide to do make sure it makes you happy.
- Nor hell a fury like a woman scorned—William Congreve
- chirp look out for them Cars coming from the Wesszzz
- Manchester derby at home revenge

The Development dataset consists of 886 Tweets with the 11 aforementioned classes and their respective labels. The class distribution in Train dataset is skewed in favor of five emotions, anger, disgust, joy, optimism, and sadness. The same class distribution is evident in the Development dataset, which is dominated by the same five emotions, Figure ??a.

SEM2018 contains 329 unique class combinations. The frequency of these unique combinations follows a power law distribution for both Train and Development datasets Figure ??b. The most frequent class combination for Train and Development was: anger and disgust, followed by joy and optimism. One third of the class combinations appears only once and often combine contradiction emotions, such as joy and sadness.

5.2. Toxic Comments

The Toxic dataset consists of two datasets as well, Train and Development. Train dataset consists of 159,571 unique comments labeled with 6 different types of toxicity: *toxic, severe_toxic, obscene, threat, insult* and *identity_hate*. Some example of comments are:

- I don't anonymously edit articles at all.
- Dear god this site is horrible.
- Please pay special attention on this
- Cant sing for s**t though.

The Development dataset consists of 63,978 toxic comments with the same classes and their respective labels. The class distribution is heavily skewed towards toxic, obscene and insult, in both Train and Development, Figure ??a.

TOXIC Train includes 29 unique class combinations, while Development has 21. Both frequencies follow a power law distribution, similar to SEM2018, Figure ??b. The most common class combinations for Train and Development are: obscene and insult, obscene only and insult only. The most uncommon class combinations included threat and severe_toxic classes. Out of 29 unique classes for Train 12 have more than 100 occurrences, while out of 21 unique class combinations of Development 7 appear more than 100 times.

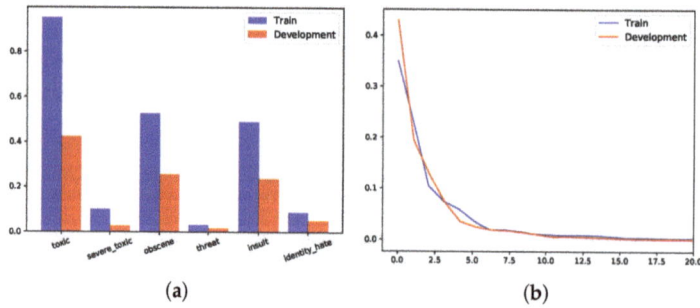

Figure 9. TOXIC Class Distribution (a) and frequency of unique class combinations (b).

6. Results

The accuracy scores are validated via 10-fold validation. The baseline neural network (NN) model [?] expectedly under-performs [?]. For the SEM2018 datasets, both ensembles outperform each individual model. The stacked ensemble provides the best results in the Train subset while the weighted ensemble marginally outperforms stacked ensemble in the Development subset, Table ??. The accuracy of both ensembles is limited, to a degree, by the inherit bias of the dataset. The performance of our ensembles outperforms all submitted models in the Codalab Competition https://competitions.codalab.org/competitions/17751#results.

Table 2. Accuracy score for SEM2018.

Model	Train	Dev.
NN	0.2946	0.1710
Model1	0.8351	0.8032
Model2	0.8352	0.8094
Model3	0.8283	0.8019
Model4	0.8385	0.8138
Model5	0.8309	0.7737
Stacked En.	**0.87**	0.8149
Weighted En.	0.8472	**0.8151**

The baseline NN performed better in TOXIC dataset. NN performance is boosted by the big number of unclassified elements in the dataset, Table ??, more than 40% of the samples as seen in Figure ??b. TOXIC dataset included more than 25,000 unique terms before cleaning. The number of unique terms affects the length of the tokenization and subsequently the dimension of the embedding. The required dimension reduction reduced the training time of each model but affected its performance. Our best performing model is in the op 35% of the Kaggle Competition submissions https://www.kaggle.com/c/jigsaw-toxic-comment-classification-challenge/leaderboard but more than 1% worse when compared to the top performing one.

Table 3. Accuracy score for TOXIC.

Model	Train	Dev.
NN	0.8981	0.9013
Model1	0.9759	0.9643
Model2	0.9637	0.9623
Model3	0.9777	0.9672
Model4	0.9776	0.968
Model5	0.9768	0.9623
Stacked En.	0.9776	0.9667
Weighted En.	**0.9782**	**0.9698**

Our ensemble methods improved upon single models in six out of eight cases. The classification accuracy is improved by at least one of the ensembles across both datasets. The ensembles for SEM2018 dataset performed excellently compared to other architectures. On the other hand, the extensive data cleaning of TOXIC -requirement due to computation/time constraints- hindered the performance of our models and their ensembles. Given the heavy class imbalance and the cleaning of TOXIC, the achieved accuracy of 97+% is decent. The baseline stacked ensemble under-performed our proposed weighted ensemble in three out of four cases.

All the models presented, and in extent their ensembles, can be further improved by a range of techniques. Test augmentation [?], hyperparameter optimization [?], bias reduction [?] and tailored emotional embeddings [? ?] are some techniques that could further improve the generalization capabilities of our networks. However, the computational load over multiple iterations is extensive, as the most complex models required hours of training per epoch and dataset.

7. Conclusions

We demonstrated that ensemble learning can improve the classification accuracy in multilabel text classification applications. We created and tested five different deep-learning architectures capable of handling multilabel binary classification tasks.

Our five DNN architectures were ensembled via two methods, stacked and weighted, and tested in two different datasets. The datasets used provide a similar multilabel classification but vary in size, term distribution and term frequency. The classification accuracy was improved by the ensemble models in both tasks. Our proposed weighted ensemble outperformed the baseline stacked ensemble in 75% of cases by 1.5% to 5.4%. Hyperparameter tuning, supervised or unsupervised, could further improve the results but with a heavy computational load, since each hyperparameter iteration requires the re-training/re-calculation of the ensemble.

Moving forward we aim to explore the creation and use of tailored emotional embeddings concatenated with word embeddings. Additionally, we are currently developing new data augmentation methods, tailored to text datasets. We are also exploring multilabel regression ensembles and architectures that could be considered to be the refinement of binary classification, whether multilabel or not.

Author Contributions: Conceptualization, G.H. and I.A.; methodology, G.H.; software, G.H.; validation, G.H.; formal analysis, G.H.; investigation, G.H.; resources, G.H.; data curation, G.H.; writing—original draft preparation, G.H. and I.A.; writing—review and editing, G.H., I.A. and D.M.; visualization, G.H.; supervision, I.A. and D.M.; funding acquisition, D.M. All authors have read and agreed to the published version of the manuscript.

Funding: This research was funded by Engineering and Physical Sciences Research Council grant number EP/M02315X/1: "From Human Data to Personal Experience".

Conflicts of Interest: The authors declare no conflict of interest.

References

- Barbieri, F.; Basile, V.; Croce, D.; Nissim, M.; Novielli, N.; Patti, V. Overview of the Evalita 2016 Sentiment Polarity Classification Task. 2016. Available online: http://ceur-ws.org/Vol-1749/paper_026.pdf (accessed on 31 March 2020).
- Tang, F.; Fu, L.; Yao, B.; Xu, W. Aspect based fine-grained sentiment analysis for online reviews. *Inf. Sci.* **2019**, *488*, 190–204. [CrossRef]
- Poria, S.; Chaturvedi, I.; Cambria, E.; Hussain, A. Convolutional MKL based multimodal emotion recognition and sentiment analysis. In Proceedings of the 2016 IEEE 16th International Conference on Data Mining (ICDM), Barcelona, Spain, 12–15 December 2016; pp. 439–448.
- Haralabopoulos, G.; Wagner, C.; McAuley, D.; Simperl, E. A multivalued emotion lexicon created and evaluated by the crowd. In Proceedings of the 2018 Fifth International Conference on Social Networks Analysis, Management and Security (SNAMS), Valencia, Spain, 15–18 October 2018; pp. 355–362.
- Haralabopoulos, G.; Simperl, E. Crowdsourcing for beyond polarity sentiment analysis a pure emotion lexicon. *arXiv* **2017**, arXiv:1710.04203.
- Go, A.; Bhayani, R.; Huang, L. Twitter sentiment classification using distant supervision. *CS224N Proj. Rep. Stanf.* **2009**, *1*, 2009.
- Tian, Y.; Galery, T.; Dulcinati, G.; Molimpakis, E.; Sun, C. Facebook sentiment: Reactions and emojis. In Proceedings of the Fifth International Workshop on Natural Language Processing for Social Media, Valencia, Spain, 3 April 2017; pp. 11–16.
- Zhang, C.; Ma, Y. *Ensemble Machine Learning: Methods and Applications*; Springer: Berlin/Heidelberg, Germany, 2012.

- Zurada, J.M. *Introduction to Artificial Neural Systems*; West Publishing Company St. Paul: Eagan, MN, USA, 1992; Volume 8.
- Glorot, X.; Bengio, Y. Understanding the difficulty of training deep feedforward neural networks. In Proceedings of the Thirteenth International Conference on Artificial Intelligence and Statistics, Sardinia, Italy, 13–15 May 2010; pp. 249–256.
- Mollahosseini, A.; Chan, D.; Mahoor, M.H. Going deeper in facial expression recognition using deep neural networks. In Proceedings of the 2016 IEEE Winter Conference on Applications of Computer Vision (WACV), Lake Placid, NY, USA, 7–9 March 2016; pp. 1–10.
- Havaei, M.; Davy, A.; Warde-Farley, D.; Biard, A.; Courville, A.; Bengio, Y.; Pal, C.; Jodoin, P.M.; Larochelle, H. Brain tumor segmentation with deep neural networks. *Med. Image Anal.* **2017**, *35*, 18–31. [CrossRef] [PubMed]
- Abbaspour-Gilandeh, Y.; Molaee, A.; Sabzi, S.; Nabipur, N.; Shamshirband, S.; Mosavi, A. A Combined Method of Image Processing and Artificial Neural Network for the Identification of 13 Iranian Rice Cultivars. *Agronomy* **2020**, *10*, 117. [CrossRef]
- Nasukawa, T.; Yi, J. Sentiment analysis: Capturing favorability using natural language processing. In Proceedings of the 2nd International Conference on Knowledge Capture, Sanibel Island, FL, USA, 23–25 October 2003; pp. 70–77.
- Yi, J.; Nasukawa, T.; Bunescu, R.; Niblack, W. Sentiment analyzer: Extracting sentiments about a given topic using natural language processing techniques. In Proceedings of the Third IEEE International Conference on Data Mining, Melbourne, FL, USA, USA, 19–22 November 2003; pp. 427–434.
- Agarwal, A.; Xie, B.; Vovsha, I.; Rambow, O.; Passonneau, R. Sentiment analysis of twitter data. In Proceedings of the Workshop on Language in Social Media (LSM 2011), Portland, OR, USA, 23 June 2011; pp. 30–38.
- Pak, A.; Paroubek, P. Twitter as a corpus for sentiment analysis and opinion mining. *LREc* **2010**, *10*, 1320–1326.
- Cvijikj, I.P.; Michahelles, F. Understanding social media marketing: A case study on topics, categories and sentiment on a Facebook brand page. In Proceedings of the 15th International Academic MindTrek Conference: Envisioning Future Media Environments, Tampere, Finland, 28–30 September 2011; pp. 175–182.
- Mullen, T.; Malouf, R. A Preliminary Investigation into Sentiment Analysis of Informal Political Discourse. In Proceedings of the AAAI Spring Symposium: Computational Approaches to Analyzing Weblogs, Stanford, CA, USA, 27–29 March 2006; pp. 159–162.
- Denecke, K.; Deng, Y. Sentiment analysis in medical settings: New opportunities and challenges. *Artif. Intell. Med.* **2015**, *64*, 17–27. [CrossRef]
- Zhang, Y.; Chen, M.; Huang, D.; Wu, D.; Li, Y. iDoctor: Personalized and professionalized medical recommendations based on hybrid matrix factorization. *Future Gener. Comput. Syst.* **2017**, *66*, 30–35. [CrossRef]
- Picard, R.W. *Affective Computing*; MIT Press: Cambridge, MA, USA, 2000.
- Cambria, E. Affective computing and sentiment analysis. *IEEE Intell. Syst.* **2016**, *31*, 102–107. [CrossRef]
- Alm, C.O.; Roth, D.; Sproat, R. Emotions from text: Machine learning for text-based emotion prediction. In Proceedings of the Conference on Human Language Technology and Empirical Methods in Natural Language Processing, Vancouver, BC, Canada, 6–8 October 2005; pp. 579–586.
- Boiy, E.; Hens, P.; Deschacht, K.; Moens, M.F. Automatic Sentiment Analysis in On-line Text. In Proceedings of the ELPUB2007 Conference on Electronic Publishing, Vienna, Austria, 13–15 June 2007; pp. 349–360.
- Asghar, M.Z.; Subhan, F.; Imran, M.; Kundi, F.M.; Shamshirband, S.; Mosavi, A.; Csiba, P.; Varkonyi-Koczy, A.R. Performance evaluation of supervised machine learning techniques for efficient detection of emotions from online content. *arXiv* **2019**, arXiv:1908.01587.
- Chaffar, S.; Inkpen, D. Using a heterogeneous dataset for emotion analysis in text. In Proceedings of the Canadian Conference on Artificial Intelligence, St. John's, NL, Canada, 25–27 May 2011; pp. 62–67.
- Sidorov, G.; Miranda-Jiménez, S.; Viveros-Jiménez, F.; Gelbukh, A.; Castro-Sánchez, N.; Velásquez, F.; Díaz-Rangel, I.; Suárez-Guerra, S.; Trevino, A.; Gordon, J. Empirical study of machine learning based approach for opinion mining in tweets. In Proceedings of the Mexican international conference on Artificial intelligence, San Luis Potosí, Mexico, 27 October–4 November 2012; pp. 1–14.
- Boiy, E.; Moens, M.F. A machine learning approach to sentiment analysis in multilingual Web texts. *Inf. Retr.* **2009**, *12*, 526–558. [CrossRef]

- Annett, M.; Kondrak, G. A comparison of sentiment analysis techniques: Polarizing movie blogs. In Proceedings of the Conference of the Canadian Society for Computational Studies of Intelligence, Windsor, ON, Canada, 28–30 May 2008; pp. 25–35.
- Dos Santos, C.; Gatti, M. Deep convolutional neural networks for sentiment analysis of short texts. In Proceedings of the 25th International Conference on Computational Linguistics (COLING 2014), Dublin, Ireland, 23–29 August 2014; pp. 69–78.
- Wang, X.; Jiang, W.; Luo, Z. Combination of convolutional and recurrent neural network for sentiment analysis of short texts. In Proceedings of the 26th International Conference on Computational Linguistics (COLING 2016), Osaka, Japan, 11–16 December 2016; pp. 2428–2437.
- Severyn, A.; Moschitti, A. Learning to rank short text pairs with convolutional deep neural networks. In Proceedings of the 38th international ACM SIGIR Conference on Research and Development in Information Retrieval, Santiago, Chile, 9–13 August 2015; pp. 373–382.
- Lai, S.; Xu, L.; Liu, K.; Zhao, J. Recurrent convolutional neural networks for text classification. In Proceedings of the Twenty-Ninth AAAI Conference on Artificial Intelligence, Austin, TX, USA, 25–30 January 2015.
- Wu, Y.; Jiang, M.; Lei, J.; Xu, H. Named entity recognition in Chinese clinical text using deep neural network. *Stud. Health Technol. Inform.* **2015**, *216*, 624. [PubMed]
- Sutskever, I.; Martens, J.; Hinton, G.E. Generating text with recurrent neural networks. In Proceedings of the 28th International Conference on Machine Learning (ICML-11), Bellevue, WC, USA, 28 June–2 July 2011; pp. 1017–1024.
- Tang, D.; Qin, B.; Liu, T. Document modeling with gated recurrent neural network for sentiment classification. In Proceedings of the 2015 Conference on Empirical Methods in Natural lAnguage Processing, Lisbon, Portugal, 17–21 September 2015; pp. 1422–1432. [CrossRef]
- Poria, S.; Peng, H.; Hussain, A.; Howard, N.; Cambria, E. Ensemble application of convolutional neural networks and multiple kernel learning for multimodal sentiment analysis. *Neurocomputing* **2017**, *261*, 217–230. [CrossRef]
- Araque, O.; Corcuera-Platas, I.; Sanchez-Rada, J.F.; Iglesias, C.A. Enhancing deep learning sentiment analysis with ensemble techniques in social applications. *Expert Syst. Appl.* **2017**, *77*, 236–246. [CrossRef]
- Xu, S.; Liang, H.; Baldwin, T. Unimelb at semeval-2016 tasks 4a and 4b: An ensemble of neural networks and a word2vec based model for sentiment classification. In Proceedings of the 10th International Workshop on Semantic Evaluation (SemEval-2016), San Diego, CA, USA, 16–17 June 2016; pp. 183–189. [CrossRef]
- Deriu, J.; Gonzenbach, M.; Uzdilli, F.; Lucchi, A.; Luca, V.D.; Jaggi, M. Swisscheese at semeval-2016 task 4: Sentiment classification using an ensemble of convolutional neural networks with distant supervision. In Proceedings of the 10th International Workshop on Semantic Evaluation, San Diego, CA, USA, 16–17 June 2016; pp. 1124–1128.
- Bachman, P.; Alsharif, O.; Precup, D. Learning with pseudo-ensembles. *Adv. Neural Inf. Process. Syst.* **2014**, *27*, 3365–3373.
- Gong, Y.; Jia, Y.; Leung, T.; Toshev, A.; Ioffe, S. Deep convolutional ranking for multilabel image annotation. *arXiv* **2013**, arXiv:1312.4894.
- Cakir, E.; Heittola, T.; Huttunen, H.; Virtanen, T. Polyphonic sound event detection using multi label deep neural networks. In Proceedings of the 2015 International Joint Conference on Neural Networks (IJCNN), Killarney, Ireland, 12–17 July 2015; pp. 1–7.
- Huang, Y.; Wang, W.; Wang, L.; Tan, T. Multi-task deep neural network for multi-label learning. In Proceedings of the 2013 IEEE International Conference on Image Processing, Melbourne, VIC, Australia, 15–18 September 2013; pp. 2897–2900.
- Chen, G.; Ye, D.; Xing, Z.; Chen, J.; Cambria, E. Ensemble application of convolutional and recurrent neural networks for multi-label text categorization. In Proceedings of the 2017 International Joint Conference on Neural Networks (IJCNN), Anchorage, AK, USA, 14–19 May 2017; pp. 2377–2383.
- Li, J.; Rao, Y.; Jin, F.; Chen, H.; Xiang, X. Multi-label maximum entropy model for social emotion classification over short text. *Neurocomputing* **2016**, *210*, 247–256. [CrossRef]
- Johnson, R.; Zhang, T. Effective use of word order for text categorization with convolutional neural networks. *arXiv* **2014**, arXiv:1412.1058.

- Zhang, Z.; Robinson, D.; Tepper, J. Detecting hate speech on twitter using a convolution-gru based deep neural network. In Proceedings of the European Semantic Web Conference, Heraklion, Greece, 3–7 June 2018; pp. 745–760.
- Zhang, X.; Zhao, J.; LeCun, Y. Character-level convolutional networks for text classification. *Adv. Neural Inf. Process. Syst.* **2015**, *28*, 649–657.
- Brownlee, J. *Deep Learning with Python: Develop Deep Learning Models on Theano and Tensorflow Using Keras*; Machine Learning Mastery: Vermont, Australia, 2016.
- Yao, Z.; Ze-wen, C. Research on the construction and filter method of stop-word list in text preprocessing. In Proceedings of the 2011 Fourth International Conference on Intelligent Computation Technology and Automation, Shenzhen, China, 28–29 March 2011; Volume 1, pp. 217–221.
- Mikolov, T.; Sutskever, I.; Chen, K.; Corrado, G.S.; Dean, J. Distributed representations of words and phrases and their compositionality. *Adv. Neural Inf. Process. Syst.* **2013**, *26*, 3111–3119.
- Pennington, J.; Socher, R.; Manning, C. Glove: Global vectors for word representation. In Proceedings of the 2014 Conference on Empirical Methods in Natural Language Processing (EMNLP), Doha, Qatar, 25–29 October 2014; pp. 1532–1543. [CrossRef]
- Srivastava, N.; Hinton, G.; Krizhevsky, A.; Sutskever, I.; Salakhutdinov, R. Dropout: A simple way to prevent neural networks from overfitting. *J. Mach. Learn. Res.* **2014**, *15*, 1929–1958.
- Livieris, I.E.; Iliadis, L.; Pintelas, P. On ensemble techniques of weight-constrained neural networks. *Evolv. Syst.* **2020**, *11*, 1–13. [CrossRef]
- Hochreiter, S.; Schmidhuber, J. Long short-term memory. *Neural Comput.* **1997**, *9*, 1735–1780. [CrossRef] [PubMed]
- Cho, K.; Van Merriënboer, B.; Bahdanau, D.; Bengio, Y. On the properties of neural machine translation: Encoder-decoder approaches. *arXiv* **2014**, arXiv:1409.1259.
- Yin, W.; Kann, K.; Yu, M.; Schütze, H. Comparative study of CNN and RNN for natural language processing. *arXiv* **2017**, arXiv:1702.01923.
- Le, H.T.; Cerisara, C.; Denis, A. Do convolutional networks need to be deep for text classification? In Proceedings of the Workshops at the Thirty-Second AAAI Conference on Artificial Intelligence, New Orleans, LA, USA, 2–7 February 2018.
- Qiu, X.; Zhang, L.; Ren, Y.; Suganthan, P.N.; Amaratunga, G. Ensemble deep learning for regression and time series forecasting. In Proceedings of the 2014 IEEE symposium on Computational Intelligence in Ensemble Learning (CIEL), Orlando, FL, USA, 9–12 December 2014; pp. 1–6.
- Sagi, O.; Rokach, L. Ensemble learning: A survey. *Wiley Interdiscip. Rev. Data Min. Knowl. Discov.* **2018**, *8*, e1249. [CrossRef]
- Ekbal, A.; Saha, S. Stacked ensemble coupled with feature selection for biomedical entity extraction. *Knowl. Based Syst.* **2013**, *46*, 22–32. [CrossRef]
- Zhai, B.; Chen, J. Development of a stacked ensemble model for forecasting and analyzing daily average PM2.5 concentrations in Beijing, China. *Sci. Total Environ.* **2018**, *635*, 644–658. [CrossRef]
- Diab, D.M.; El Hindi, K.M. Using differential evolution for fine tuning naïve Bayesian classifiers and its application for text classification. *Appl. Soft Comput.* **2017**, *54*, 183–199. [CrossRef]
- Zhang, Y.; Liu, B.; Cai, J.; Zhang, S. Ensemble weighted extreme learning machine for imbalanced data classification based on differential evolution. *Neural Comput. Appl.* **2017**, *28*, 259–267. [CrossRef]
- Storn, R.; Price, K. Differential evolution—A simple and efficient heuristic for global optimization over continuous spaces. *J. Glob. Optim.* **1997**, *11*, 341–359. [CrossRef]
- SUTER, B.W. The multilayer perceptron as an approximation to a Bayes optimal discriminant function. *IEEE Trans. Neural Networks* **1990**, *1*, 291.
- Koutsoukas, A.; Monaghan, K.J.; Li, X.; Huan, J. Deep-learning: Investigating deep neural networks hyper-parameters and comparison of performance to shallow methods for modeling bioactivity data. *J. Cheminf.* **2017**, *9*, 42. [CrossRef] [PubMed]
- Zhang, X.; LeCun, Y. Text understanding from scratch. *arXiv* **2015**, arXiv:1502.01710.
- Loshchilov, I.; Hutter, F. CMA-ES for hyperparameter optimization of deep neural networks. *arXiv* **2016**, arXiv:1604.07269.

- Tao, Y.; Gao, X.; Hsu, K.; Sorooshian, S.; Ihler, A. A deep neural network modeling framework to reduce bias in satellite precipitation products. *J. Hydrometeorol.* **2016**, *17*, 931–945. [CrossRef]
- Haralabopoulos, G.; Wagner, C.; McAuley, D.; Anagnostopoulos, I. Paid Crowdsourcing, Low Income Contributors, and Subjectivity. In Proceedings of the IFIP International Conference on Artificial Intelligence Applications and Innovations, Crete, Greece, 24–26 May 2019; pp. 225–231.

© 2020 by the authors. Licensee MDPI, Basel, Switzerland. This article is an open access article distributed under the terms and conditions of the Creative Commons Attribution (CC BY) license (http://creativecommons.org/licenses/by/4.0/).

Article

Ensemble Deep Learning Models for Forecasting Cryptocurrency Time-Series

Ioannis E. Livieris [1,*], Emmanuel Pintelas [1], Stavros Stavroyiannis [2] and Panagiotis Pintelas [1]

1 Department of Mathematics, University of Patras, GR 265-00 Patras, Greece; ece6835@upnet.gr (E.P.); ppintelas@gmail.com (P.P.)
2 Department of Accounting & Finance, University of the Peloponnese, GR 241-00 Antikalamos, Greece; computmath@gmail.com
* Correspondence: livieris@upatras.gr

Received: 17 April 2020; Accepted: 8 May 2020; Published: 10 May 2020

Abstract: Nowadays, cryptocurrency has infiltrated almost all financial transactions; thus, it is generally recognized as an alternative method for paying and exchanging currency. Cryptocurrency trade constitutes a constantly increasing financial market and a promising type of profitable investment; however, it is characterized by high volatility and strong fluctuations of prices over time. Therefore, the development of an intelligent forecasting model is considered essential for portfolio optimization and decision making. The main contribution of this research is the combination of three of the most widely employed ensemble learning strategies: ensemble-averaging, bagging and stacking with advanced deep learning models for forecasting major cryptocurrency hourly prices. The proposed ensemble models were evaluated utilizing state-of-the-art deep learning models as component learners, which were comprised by combinations of long short-term memory (LSTM), Bi-directional LSTM and convolutional layers. The ensemble models were evaluated on prediction of the cryptocurrency price on the following hour (regression) and also on the prediction if the price on the following hour will increase or decrease with respect to the current price (classification). Additionally, the reliability of each forecasting model and the efficiency of its predictions is evaluated by examining for autocorrelation of the errors. Our detailed experimental analysis indicates that ensemble learning and deep learning can be efficiently beneficial to each other, for developing strong, stable, and reliable forecasting models.

Keywords: deep learning; ensemble learning; convolutional networks; long short-term memory; cryptocurrency; time-series

1. Introduction

The global financial crisis of 2007–2009 was the most severe crisis over the last few decades with, according to the National Bureau of Economic Research, a peak to trough contraction of 18 months. The consequences were severe in most aspects of life including economy (investment, productivity, jobs, and real income), social (inequality, poverty, and social tensions), leading in the long run to political instability and the need for further economic reforms. In an attempt to "think outside the box" and bypass the governments and financial institutions manipulation and control, Satoshi Nakamoto [1] proposed Bitcoin which is an electronic cash allowing online payments, where the double-spending problem was elegantly solved using a novel purely peer-to-peer decentralized blockchain along with a cryptographic hash function as a proof-of-work.

Nowadays, there are over 5000 cryptocurrencies available; however, when it comes to scientific research there are several issues to deal with. The large majority of these are relatively new, indicating that there is an insufficient amount of data for quantitative modeling or price forecasting. In the same manner, they are not highly ranked when it comes to market capitalization to be considered as market drivers. A third aspect which has not attracted attention in the literature is the separation of cryptocurrencies between mineable and non-mineable. Minable cryptocurrencies have several advantages i.e., the performance of different mineable coins can be monitored within the same blockchain which cannot be easily said for non-mineable coins, and they are community driven open source where different developers can contribute, ensuring the fact that a consensus has to be reached before any major update is done, in order to avoid splitting. Finally, when it comes to the top cryptocurrencies, it appears that mineable cryptocurrencies like Bitcoin (BTC) and Ethereum (ETH), recovered better the 2018 crash rather than Ripple (XRP) which is the highest ranked pre-mined coin. In addition, the non-mineable coins transactions are powered via a centralized blockchain, endangering price manipulation through inside trading, since the creators keep a given percentage to themselves, or through the use of pump and pull market mechanisms. Looking at the number one cryptocurrency exchange in the world, Coinmarketcap, by January 2020 at the time of writing there are only 31 mineable cryptocurrencies out of the first 100, ranked by market capitalization. The classical investing strategy in cryptocurrency market is the "buy, hold and sell" strategy, in which cryptocurrencies are bought with real money and held until reaching a higher price worth selling in order for an investor to make a profit. Obviously, a potential fractional change in the price of a cryptocurrency may gain opportunities for huge benefits or significant investment losses. Thus, the accurate prediction of cryptocurrency prices can potentially assist financial investors for making their proper investment policies by decreasing their risks. However, the accurate prediction of cryptocurrency prices is generally considered a significantly complex and challenging task, mainly due to its chaotic nature. This problem is traditionally addressed by the investor's personal experience and consistent watching of exchanges prices. Recently, the utilization of intelligent decision systems based on complicated mathematical formulas and methods have been adopted for potentially assisting investors and portfolio optimization.

Let $y_1, y_2, \ldots, y_n$ be the observations of a time series. Generally, a nonlinear regression model of order m is defined by

$$y_t = f(y_{t-1}, y_{t-2}, \ldots, y_{t-m}, \theta), \tag{1}$$

where m values of y_t, θ is the parameter vector. After the model structure has been defined, function $f(\cdot)$ can be determined by traditional time-series methods such as ARIMA (Auto-Regressive Integrated Moving Average) and GARCH-type models and their variations [2–4] or by machine learning methods such as Artificial Neural Networks (ANNs) [5,6]. However, both mentioned approaches fail to depict the stochastic and chaotic nature of cryptocurrency time-series and be successfully effective for accurate forecasting [7]. To this end, more sophisticated algorithmic approaches have to be applied such as deep learning and ensemble learning methods. From the perspective of developing strong forecasting models, deep learning and ensemble learning constitute two fundamental learning strategies. The former is based on neural networks architectures and it is able to achieve state-of-the-art accuracy by creating and exploiting new more valuable features by filtering out the noise of the input data; while the latter attempts to generate strong prediction models by exploiting multiple learners in order to reduce the bias or variance of error.

During the last few years, researchers paid special attention to the development of time-series forecasting models which exploit the advantages and benefits of deep learning techniques such as convolutional and long short-term memory (LSTM) layers. More specifically, Wen and Yuan [8] and Liu et al. [9] proposed Convolutional Neural Network (CNN) and LSTM prediction models for stock market forecasting. Along this line, Livieris et al. [10] and Pintelas et al. [11] proposed CNN-LSTM models with various architectures for efficiently forecasting gold and cryptocurrency time-series price and movement, reporting some interesting results. Nevertheless, although deep learning models are tailored to cope with temporal correlations and efficiently extract more valuable information from the training set, they failed to generate reliable forecasting models [7,11]; while in contrast ensemble learning models although they are an elegant solution to develop stable models and address the high variance of individual forecasting models, their performance heavily depends on the diversity and accuracy of the component learners. Therefore, a time-series prediction model, which exploits the benefits of both mentioned methodologies may significantly improve the prediction performance.

The main contribution of this research is the combination of ensemble learning strategies with advanced deep learning models for forecasting cryptocurrency hourly prices and movement. The proposed ensemble models utilize state-of-the-art deep learning models as component learners which are based on combinations of Long Short-Term Memory (LSTM), Bi-directional LSTM (BiLSTM) and convolutional layers. An extensive experimental analysis is performed considering both classification and regression problems, to evaluate the performance of averaging, bagging and stacking ensemble strategies. More analytically, all ensemble models are evaluated on prediction of the cryptocurrency price on the next hour (regression) and also on the prediction if the price on the following hour will increase or decrease with respect to the current price (classification). It is worth mentioning that the information of predicting the movement of a cryptocurrency is probably more significant that the prediction of the price for investors and financial institutions. Additionally, the efficiency of the predictions of each forecasting model is evaluated by examining for autocorrelation of the errors, which constitutes a significant reliability test of each model.

The remainder of the paper is organized as follows: Section 2 presents a brief review of state of the art deep learning methodologies for cryptocurrency forecasting. Section 3 presents the advanced deep learning models, while Section 4 presents the ensemble strategies utilized in our research. Section 5 presents our experimental methodology including the data preparation and preprocessing as well as the detailed experimental analysis, regarding the evaluation of ensemble of deep learning models. Section 6 discusses the obtained results and summarizes our findings. Finally, Section 7 presents our conclusions and presents some future directions.

2. Deep Learning in Cryptocurrency Forecasting: State-of-the-Art

Yiying and Yeze [12] focused on the price non-stationary dynamics of three cryptocurrencies Bitcoin, Etherium, and Ripple. Their approach aimed at identifying and understand the factors which influence the value formation of these digital currencies. Their collected data contained 1030 trading days regarding opening, high, low, and closing prices. They conducted an experimental analysis which revealed the efficiency of LSTM models over classical ANNs, indicating that LSTM models are more capable of exploiting information hidden in historical data. Additionally, the authors stated that probably the reason for the efficiency of LSTM networks is that they tend to depend more on short-term dynamics while ANNs tends to depend more on long-term history. Nevertheless, in case enough historical information is given, ANNs can achieve similar accuracy to LSTM networks.

Nakano et al. [13] examined the performance of ANNs for the prediction of Bitcoin intraday technical trading. The authors focused on identifying the key factors which affect the prediction performance for extracting useful trading signals of Bitcoin from its technical indicators. For this purposed, they conducted a series of experiments utilizing various ANN models with shallow and deep architectures and datasets structures The data utilized in their research regarded Bitcoin time-series return data at 15-min time intervals. Their experiments illustrated that the utilization of multiple technical indicators could possibly prevent the prediction model from overfitting of non-stationary financial data, which enhances trading performance. Moreover, they stated that their proposed methodology attained considerably better performance than the primitive technical trading and buy-and-hold strategies, under realistic assumptions of execution costs.

Mcnally et al. [14] utilized two deep learning models, namely a Bayesian-optimised Recurrent Neural Network and a LSTM network, for Bitcoin price prediction. The utilized data ranged from the August 2013 until July 2016, regarding open, high, low and close of Bitcoin prices as well as the block difficulty and hash rate. Their performance evaluation showed that the LSTM network demonstrated the best prediction accuracy, outperforming the other recurrent model as well as the classical statistical method ARIMA.

Shintate and Pichl [15] proposed a new trend prediction classification framework which is based on deep learning techniques. Their proposed framework utilized a metric learning-based method, called Random Sampling method, which measures the similarity between the training samples and the input patterns. They used high frequency data (1-min) ranged from June 2013 to March 2017 containing historical data from OkCoin Bitcoin market (Chinese Yuan Renminbi and US Dollars). The authors concluded that the profit rates based on utilized sampling method considerably outperformed those based on LSTM networks, confirming the superiority of the proposed framework. In contrast, these profit rates were lower than those obtained of the classical buy-and-hold strategy; thus they stated that it does not provide a basis for trading.

Miura et al. [16] attempted to analyze the high-frequency Bitcoin (1-min) time series utilizing machine learning and statistical forecasting models. Due to the large size of the data, they decided to aggregate the realized volatility values utilizing 3-h long intervals. Additionally, they pointed out that these values presented a weak correlation based on high-low price extent with the relative values of the 3-h interval. In their experimental analysis, they focused on evaluating various ANNs-type models, SVMs and Ridge Regression and the Heterogeneous Auto-Regressive Realized Volatility model. Their results demonstrated that Ridge Regression considerably presented the best performance while SVM exhibited poor performance.

Ji et al. [17] evaluated the prediction performance on Bitcoin price of various deep learning models such as LSTM networks, convolutional neural networks, deep neural networks, deep residual networks and their combinations. The data used in their research, contained 29 features of the Bitcoin blockchain from 2590 days (from 29 November 2011 to 31 December 2018). They conducted a detailed experimental procedure considering both classification and regression problems, where the former predicts whether or not the next day price will increase or decrease and the latter predicts the next day's Bitcoin price. The numerical experiments illustrated that the deep neural DNN-based models performed the best for price ups-and-downs while the LSTM models slightly outperformed the rest of the models for forecasting Bitcoin's price.

Kumar and Rath [18] focused on forecasting the trends of Etherium prices utilizing machine learning and deep learning methodologies. They conducted an experimental analysis and compared the prediction ability of LSTM neural networks and Multi-Layer perceptron (MLP). They utilized daily, hourly, and minute based data which were collected from the CoinMarket and CoinDesk repositories. Their evaluation results illustrated that LSTM marginally outperformed MLP but not considerably, although their training time was significantly high.

Pintelas et al. [7,11] conducted a detailed research, evaluating advanced deep learning models for predicting major cryptocurrency prices and movements. Additionally, they conducted a detailed discussion regarding the fundamental research questions: Can deep learning algorithms efficiently predict cryptocurrency prices? Are cryptocurrency prices a random walk process? Which is a proper validation method of cryptocurrency price prediction models? Their comprehensive experimental results revealed that even the LSTM-based and CNN-based models, which are generally preferable for time-series forecasting [8–10], were unable to generate efficient and reliable forecasting models. Moreover, the authors stated that cryptocurrency prices probably follow an almost random walk process while few hidden patterns may probably exist. Therefore, new sophisticated algorithmic approaches should be considered and explored for the development of a prediction model to make accurate and reliable forecasts.

In this work, we advocate combining the advantages of ensemble learning and deep learning for forecasting cryptocurrency prices and movement. Our research contribution aims on exploiting the ability of deep learning models to learn the internal representation of the cryptocurrency data and the effectiveness of ensemble learning for generating powerful forecasting models by exploiting multiple learners for reducing the bias or variance of error. Furthermore, similar to our previous research [7,11], we provide detailed performance evaluation for both regression and classification problems. To the best of our knowledge, this is the first research devoted to the adoption and combination of ensemble learning and deep learning for forecasting cryptocurrencies prices and movement.

3. Advanced Deep Learning Techniques

3.1. Long Short-Term Memory Neural Networks

Long Short-term memory (LSTM) [19] constitutes a special case of recurrent neural networks which were originally proposed to model both short-term and long-term dependencies [20–22]. The major novelty unit in a LSTM network is the memory block in the recurrent hidden layer which contains memory cells with self-connections memorizing the temporal state and adaptive gate units for controlling the information flow in the block. With the treatment of the hidden layer as a memory unit, LSTM can cope the correlation within time-series in both short and long term [23].

More analytically, the structure of the memory cell c_t is composed by three gates: the input gate, the forget gate and the output gate. At every time step t, the input gate i_t determines which information is added to the cell state S_t (memory), the forget gate f_t determines which information is thrown away from the cell state through the decision by a transformation function in the forget gate layer; while the output gate o_t determines which information from the cell state will be used as output.

With the utilization of gates in each cell, data can be filtered, discarded or added. In this way, LSTM networks are capable of identifying both short and long term correlation features within time series. Additionally, it is worth mentioning that a significant advantage of the utilization of memory cells and adaptive gates which control information flow is that the vanishing gradient problem can be considerably addressed, which is crucial for the generalization performance of the network [20].

The simplest way to increase the depth and the capacity of LSTM networks is to stack LSTM layers together, in which the output of the $(L-1)$th LSTM layer at time t is treated as input of the Lth layer. Notice that this input-output connections are the only connections between the LSTM layers of the network. Based on the above formulation, the structure of the stacked LSTM can be describe as follows: Let h_t^L and h_t^{L-1} denote outputs in the Lth and $(L-1)$th layer, respectively. Each layer L produces a hidden state h_t^L based on the current output of the previous layer h_t^{L-1} and time h_{t-1}^L. More specifically, the forget gate f_t^L of the L layer calculates the input for cell state c_{t-1}^L by

$$f_t = \sigma\left(W_f^L[h_{t-1}^L, h_t^{L-1}] + b_f^L\right),$$

where $\sigma(\cdot)$ is a sigmoid function while W_f^L and b_f are the weights matrix and bias vector of layer L regarding the forget gate, respectively. Subsequently, the input gate i_t^L of the L layer computes the values to be added to the memory cell c_t^L by

$$i_t^L = \sigma\left(W_i^L[h_{t-1}^L, h_t^{L-1}] + b_i^L\right),$$

where W_i^L is the weights matrix of layer L regarding the input gate. Then, the output gate o_t^L of the Lth layer filter the information and calculated the output value by

$$o_t^L = \sigma\left(W_o^L[h_{t-1}^L, h_t^{L-1}] + b_o^L\right),$$

where W_f^o and b_o are the weights matrix and bias vector of the output gate in the L layer, respectively. Finally, the output of the memory cell is computed by

$$h_t = o_t^L \cdot \tanh\left(c_t^L\right),$$

where $\cdot$ denotes the pointwise vector multiplication, tanh the hyperbolic tangent function and

$$c_t^L = f_t^L \cdot c_{t-1}^L + i_t^L \cdot \tilde{c}_{t-1}^L,$$
$$\tilde{c}_t^L = \tanh\left(W_{\tilde{c}}^L [h_{t-1}^L, h_t^{L-1}] + b_{\tilde{c}}^L\right).$$

3.2. Bi-Directional Recurrent Neural Networks

Similar with the LSTM networks, one of the most efficient and widely utilized RNNs architectures are the Bi-directional Recurrent Neural Networks (BRNNs) [24]. In contrast with the LSTM, these networks are composed by two hidden layers, connected to input and output. The principle idea of BRNNs is that each training sequence is presented forwards and backwards into two separate recurrent networks [20]. More specifically, the first hidden layer possesses recurrent connections from the past time steps, while in the second one, the recurrent connections are reversed, transferring activation backwards along the sequence. Given the input and target sequences, the BRNN can be unfolded across time in order to be efficiently trained utilizing a classical backpropagation algorithm.

In fact, BRNN and LSTM are based on compatible techniques in which the former proposes the wiring of two hidden layers, which compose the network, while the latter proposes a new fundamental unit for composing the hidden layer.

Along this line, Bi-directional LSTM (BiLSTM) networks [25] were proposed in the literature, which incorporate two LSTM networks in the BRNN framework. More specifically, BiLSTM incorporates a forward LSTM layer and a backward LSTM layer in order to learn information from preceding and following tokens. In this way, both past and future contexts for a given time t are accessed, hence better prediction can be achieved by taking advantage of more sentence-level information.

In a bi-directional stacked LSTM network, the output of each feed-forward LSTM layer is the same as in the classical stacked LSTM layer and these layers are iterated from $t = 1$ to T. In contrast, the output of each backward LSTM layer is iterated reversely, i.e., from $t = T$ to 1. Hence, at time t, the output of value $\overleftarrow{h}_t$ in backward LSTM layer L can be calculated as follows

$$\overleftarrow{f}_t^L = \sigma\left(W_{\overleftarrow{f}}^L [h_{t-1}^L, h_t^{L-1}] + b_{\overleftarrow{f}}^L\right),$$
$$\overleftarrow{i}_t^L = \sigma\left(W_{\overleftarrow{i}}^L [h_{t-1}^L, h_t^{L-1}] + b_{\overleftarrow{i}}^L\right),$$
$$\overleftarrow{o}_t^L = \sigma\left(W_{\overleftarrow{o}}^L [h_{t-1}^L, h_t^{L-1}] + b_{\overleftarrow{o}}^L\right),$$
$$\overleftarrow{c}_t^L = \overleftarrow{f}_t^L \cdot \overleftarrow{c}_{t-1}^L + \overleftarrow{i}_t^L \cdot \overleftarrow{\tilde{c}}_{t-1}^L,$$
$$\overleftarrow{\tilde{c}}_t^L = \tanh\left(W_{\overleftarrow{\tilde{c}}}^L [h_{t-1}^L, h_t^{L-1}] + b_{\overleftarrow{\tilde{c}}}^L\right),$$
$$\overleftarrow{h}_t^L = \overleftarrow{o}_t^L \cdot \tanh\left(\overleftarrow{c}_t^L\right).$$

Finally, the output of this BiLSTM architecture is given by

$$y_t = W\left[\overrightarrow{h}_t, \overleftarrow{h}_t\right] + b.$$

3.3. Convolutional Neural Networks

Convolutional Neural Network (CNN) models [26,27] were originally proposed for image recognition problems, achieving human level performance in many cases. CNNs have great potential to identify the complex patterns hidden in time series data. The advantage of the utilization of CNNs for time series is that they can efficiently extract knowledge and learn an internal representation from the raw time series data directly and they do not require special knowledge from the application domain to filter input features [10].

A typical CNN consists of two main components: In the first component, mathematical operations, called convolution and pooling, are utilized to develop features of the input data while in the second component, the generated features are used as input to a usually fully-connected neural network.

The convolutional layer constitutes the core of a CNN which systematically applies trained filters to input data for generating feature maps. Convolution can be considered as applying and sliding a one dimension (time) filter over the time series [28]. Moreover, since the output of a convolution is a new filtered time series, the application of several convolutions implies the generation of a multivariate times series whose dimension equals with the number of utilized filters in the layer. The rationale behind this strategy is that the application of several convolution leads to the generation of multiple discriminative features which usually improve the model's performance. In practice, this kind of layer is proven to be very efficient and stacking different convolutional layers allows deeper layers to learn high-order or more abstract features and layers close to the input to learn low-level features.

Pooling layers were proposed to address the limitation that feature maps generated by the convolutional layers, record the precise position of features in the input. These layers aggregate over a sliding window over these feature maps, reducing their length, in order to attain some translation invariance of the trained features. More analytically, the feature maps obtained from the previous convolutional layer are pooled over temporal neighborhood separately by sum pooling function or by max pooling function in order to developed a new set of pooled feature maps. Notice that the output pooled feature maps constitute a filtered version of the features maps which are imported as inputs in the pooling layer [28]. This implies that small translations of the inputs of the CNN, which are usually detected by the convolutional layers, will become approximately invariant.

Finally, in addition to convolutional and pooling layers, some include batch normalization layers [29] and dropout layers [30] in order to accelerate the training process and reduce overfitting, respectively.

4. Ensemble Deep Learning Models

Ensemble learning has been proposed as an elegant solution to address the high variance of individual forecasting models and reduce the generalization error [31–33]. The basic principle behind any ensemble strategy is to weigh a number of models and combine their individual predictions for improving the forecasting performance; while the key point for the effectiveness of the ensemble is that its components should be characterized by accuracy and diversity in their predictions [34]. In general, the combination of multiple models predictions adds a bias which in turn counters the variance of a single trained model. Therefore, by reducing the variance in the predictions, the ensemble can perform better than any single best model.

In the literature, several strategies were proposed to design and develop ensemble of regression models. Next, we present three of the most efficient and widely employed strategies: ensemble-averaging, bagging, and stacking.

4.1. Ensemble-Averaging of Deep Learning Models

Ensemble-averaging [35] (or averaging) is the simplest combination strategy for exploiting the prediction of different regression models. It constitutes a commonly and widely utilized ensemble strategy of individual trained models in which their predictions are treated equally. More specifically, each forecasting model is individually trained and the ensemble-averaging strategy linearly combines all predictions by averaging them to develop the output. Figure 1 illustrates a high-level schematic representation of the ensemble-averaging of deep learning models.

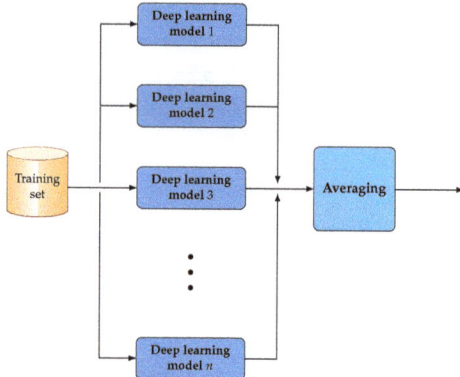

Figure 1. Ensemble-averaging of deep learning models.

Ensemble-averaging is based on the philosophy that its component models will not usually make the same error on new unseen data [36]. In this way, the ensemble model reduces the variance in the prediction, which results in better predictions compared to a single model. The advantages of this strategy are its simplicity of implementation and the exploitation of the diversity of errors of its component models without requiring any additional training on large quantities of the individual predictions.

4.2. Bagging Ensemble of Deep Learning Models

Bagging [33] is one the most widely used and successful ensemble strategies for improving the forecasting performance of unstable models. Its basic principle is the development of more diverse forecasting models by modifying the distribution of the training set based on a stochastic strategy. More specifically, it applies the same learning algorithm on different bootstrap samples of the original training set and the final output is produced via a simple averaging. An attractive property of the bagging strategy is that it reduces variance while simultaneously retains the bias which assists in avoiding overfitting [37,38]. Figure 2 demonstrates a high-level schematic representation of the bagging ensemble of n deep learning models.

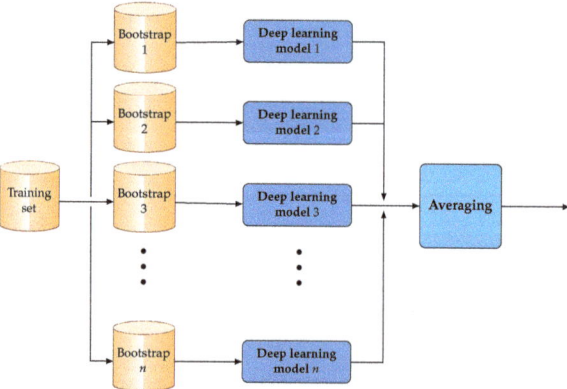

Figure 2. Bagging ensemble of deep learning models.

It is worth mentioning that bagging strategy is significantly useful for dealing with large and high-dimensional datasets where finding a single model which can exhibit good performance in one step is impossible due to the complexity and scale of the prediction problem.

4.3. Stacking Ensemble of Deep Learning Models

Stacked generalization or stacking [39] constitutes a more elegant and sophisticated approach for combining the prediction of different learning models. The motivation of this approach is based on the limitation of simple ensemble-average which is that each model is equally considered to the ensemble prediction, regardless of how well it performed. Instead, stacking induces a higher-level model for exploiting and combining the prediction of the ensemble's component models. More specifically, the models which comprise the ensemble (Level-0 models) are individually trained using the same training set (Level-0 training set). Subsequently, a Level-1 training set is generated by the collected outputs of the component classifiers.

This dataset is utilized to train a single Level-1 model (meta-model) which ultimately determines how the outputs of the Level-0 models should be efficiently combined, to maximize the forecasting performance of the ensemble. Figure 3 illustrates the stacking of deep learning models.

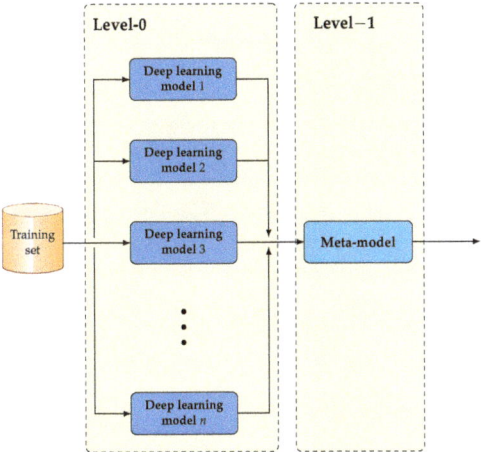

Figure 3. Stacking ensemble of deep learning models.

In general, stacked generalization works by deducing the biases of the individual learners with respect to the training set [33]. This deduction is performed by the meta-model. In other words, the meta-model is a special case of weighted-averaging, which utilizes the set of predictions as a context and conditionally decides to weight the input predictions, potentially resulting in better forecasting performance.

5. Numerical Experiments

In this section, we evaluate the performance of the three presented ensemble strategies which utilize advanced deep learning models as component learners. The implementation code was written in Python 3.4 while for all deep learning models Keras library [40] was utilized and Theano as back-end.

For the purpose of this research, we utilized data from 1 January 2018 to 31 August 2019 from the hourly price of the cryptocurrencies BTC, ETH and XRP. For evaluation purposes, the data were divided in training set and in testing set as in [7,11]. More specifically, the training set comprised data from 1 January 2018 to 28 February 2019 (10,177 datapoints), covering a wide range of long and short term trends while the testing set consisted of data from 1 March 2019 to 31 August 2019 (4415 datapoints) which ensured a substantial amount of unseen out-of-sample prices for testing.

Next, we concentrated on the experimental analysis to evaluate the presented ensemble strategies using the advanced deep learning models CNN-LSTM and CNN-BiLSTM as base learners. A detailed description of both component models is presented in Table 1. These models and their hyper-parameters were selected in previous research [7] after extensive experimentation, in which they exhibited the best performance on the utilized datasets. Both component models were trained for 50 epochs with Adaptive Moment Estimation (ADAM) algorithm [41] with a batch size equal to 512, using a mean-squared loss function. ADAM algorithm ensures that the learning steps, during the training process, are scale invariant relative to the parameter gradients.

Table 1. Parameter specification of two base learners.

Model	Description
CNN-LSTM	Convolutional layer with 32 filters of size (2,) and padding = "same". Convolutional layer with 64 filters of size (2,) and padding = "same". Max pooling layer with size (2,) LSTM layer with 70 units. Fully connected layer with 16 neurons. Output layer with 1 neuron.
CNN-BiLSTM	Convolutional layer with 64 filters of size (2,) and padding = "same". Convolutional layer with 128 filters of size (2,) and padding = "same". Max pooling layer with size (2,). BiLSTM layer with 2 × 70 units. Fully connected layer with 16 neurons. Output layer with 1 neuron.

The performance of all ensemble models was evaluated utilizing the performance metric: Root Mean Square Error (RMSE). Additionally, the classification accuracy of all ensemble deep models was measured, relative to the problem of predicting whether the cryptocurrency price would increase or decrease on the next day. More analytically, by analyzing a number of previous hourly prices, the model predicts the price on the following hour and also predicts if the price will increase or decrease, with respect to current cryptocurrency price. For this binary classification problem, three performance metrics were used: Accuracy (Acc), Area Under Curve (AUC) and F_1-score (F_1).

All ensemble models were evaluated using 7 and 11 component learners which reported the best overall performance. Notice that any attempt to increase the number of classifiers resulted to no improvement to the performance of each model. Moreover, stacking was evaluated using the most widely used state-of-the-art algorithms [42] as meta-learners: Support Vector Regression (SVR) [43], Linear Regression (LR) [44], k-Nearest Neighbor (kNN) [45] and Decision Tree Regression (DTR) [46]. For fairness and for performing an objective comparison, the hyper-parameters of all meta-learners were selected in order to maximize their experimental performance and are briefly presented in Table 2.

Table 2. Parameter specification of state-of-art algorithms used as meta-learners.

Model	Hyper-Parameters
DTR	Criterion = 'mse', max depth = unlimited, min samples split = 2.
LR	No parameters specified.
SVR	Kernel = RBF, Regularization parameter $C = 1.0$, gamma = 10^{-1}.
kNN	Number of neighbors = 10, Euclidean distance.

Summarizing, we evaluate the performance of the following ensemble models:

- "Averaging$_7$" and "Averaging$_{11}$" stand for ensemble-averaging model utilizing 7 and 11 component learners, respectively.
- "Bagging$_7$" and "Bagging$_{11}$" stand for bagging ensemble model utilizing 7 and 11 component learners, respectively.
- "Stacking$_7^{(DTR)}$" and "Stacking$_{11}^{(DTR)}$" stand for stacking ensemble model utilizing 7 and 11 component learners, respectively and DTR model as meta-learner.
- "Stacking$_7^{(LR)}$" and "Stacking$_{11}^{(LR)}$" stand for stacking ensemble model utilizing 7 and 11 component learners, respectively and LR model as meta-learner.
- "Stacking$_7^{(SVR)}$" and "Stacking$_{11}^{(SVR)}$" stand for stacking ensemble model utilizing 7 and 11 component learners, respectively and SVR as meta-learner.
- "Stacking$_7^{(kNN)}$" and "Stacking$_{11}^{(kNN)}$" stand for stacking ensemble model utilizing 7 and 11 component learners, respectively and kNN as meta-learner.

Tables 3 and 4 summarize the performance of all ensemble models using CNN-LSTM as base learner for $m = 4$ and $m = 9$, respectively. Stacking using LR as meta-learner exhibited the best regression performance, reporting the lowest RMSE score, for all cryptocurrencies. Notice that stacking$_7^{(LR)}$ and Stacking$_{11}^{(LR)}$ reported the same performance, which implies that the increment of component learners from 7 to 11 did not affect the regression performance of this ensemble algorithm. In contrast, stacking$_7^{(LR)}$ exhibited better classification performance than Stacking$_{11}^{(LR)}$, reporting higher accuracy, AUC and F_1-score. Additionally, the stacking ensemble reported the worst performance utilizing DTR and SVR as meta-learners among all ensemble models, also reporting worst performance than CNN-LSTM model; while the best classification performance was reported using kNN as meta-learner in almost all cases.

The average and bagging ensemble reported slightly better regression performance, compared to the single model CNN-LSTM. In contrast, both ensembles presented the best classification performance, considerably outperforming all other forecasting models, regarding all datasets. Moreover, the bagging ensemble reported the highest accuracy, AUC and F_1 in most cases, slightly outperforming the average-ensemble model. Finally, it is worth noticing that both the bagging and average-ensemble did not improve their performance when the number of component classifiers increased. for $m = 4$ while for $m = 9$ a slightly improvement in their performance was noticed.

Table 3. Performance of ensemble models using convolutional neural network and long short-term memory (CNN-LSTM) as base learner for $m = 4$.

Model	BTC				ETH				XRP			
	RMSE	Acc	AUC	F_1	RMSE	Acc	AUC	F_1	RMSE	Acc	AUC	F_1
CNN-LSTM	0.0101	52.80%	0.521	0.530	0.0111	53.39%	0.533	0.525	0.0110	53.82%	0.510	0.530
Averaging$_7$	0.0096	54.39%	0.543	0.536	0.0107	53.73%	0.539	**0.536**	0.0108	**54.12%**	0.552	**0.536**
Bagging$_7$	0.0097	**54.52%**	0.549	0.545	0.0107	53.95%	**0.541**	0.530	0.0095	53.29%	**0.560**	0.483
Stacking$_7^{(DTR)}$	0.0168	49.96%	0.510	0.498	0.0230	50.13%	0.506	0.497	0.0230	49.67%	0.530	0.492
Stacking$_7^{(LR)}$	**0.0085**	52.57%	0.530	0.524	**0.0093**	52.44%	0.513	0.515	**0.0095**	52.32%	0.537	0.514
Stacking$_7^{(SVR)}$	0.0128	49.82%	0.523	0.497	0.0145	51.24%	0.501	0.508	0.0146	51.27%	0.515	0.508
Stacking$_7^{(kNN)}$	0.0113	52.64%	0.525	0.511	0.0126	52.85%	0.537	0.473	0.0101	52.87%	0.536	0.514
Averaging$_{11}$	0.0096	54.16%	0.545	0.534	0.0107	53.77%	0.542	0.535	0.0107	54.12%	0.531	**0.537**
Bagging$_{11}$	0.0098	**54.31%**	0.546	**0.542**	0.0107	53.95%	**0.548**	**0.536**	0.0094	**54.32%**	0.532	0.515
Stacking$_{11}^{(DTR)}$	0.0169	49.66%	0.510	0.495	0.0224	50.42%	0.512	0.500	0.0229	50.63%	0.519	0.501
Stacking$_{11}^{(LR)}$	**0.0085**	51.05%	0.529	0.475	**0.0093**	51.95%	0.514	0.514	**0.0094**	52.35%	0.529	0.512
Stacking$_{11}^{(SVR)}$	0.0128	50.03%	0.520	0.499	0.0145	51.10%	0.510	0.506	0.0145	51.11%	0.511	0.506
Stacking$_{11}^{(kNN)}$	0.0112	52.66%	0.526	0.511	0.0126	52.65%	0.538	0.481	0.0101	53.05%	0.530	0.512

Table 4. Performance of ensemble models using CNN-LSTM as base learner for $m = 9$.

Model	BTC				ETH				XRP			
	RMSE	Acc	AUC	F_1	RMSE	Acc	AUC	F_1	RMSE	Acc	AUC	F_1
CNN-LSTM	0.0124	51.19%	0.511	0.504	0.0140	53.49%	0.535	0.513	0.0140	53.43%	0.510	0.510
Averaging$_7$	0.0122	54.12%	**0.549**	0.541	0.0138	53.86%	0.539	0.538	0.0138	**53.94%**	0.502	**0.538**
Bagging$_7$	0.0123	**54.29%**	**0.549**	0.543	0.0132	**54.05%**	**0.541**	**0.539**	0.0093	52.97%	0.501	0.502
Stacking$_7^{(DTR)}$	0.0213	50.57%	0.510	0.504	0.0280	51.58%	0.509	0.511	0.0279	51.87%	0.514	0.514
Stacking$_7^{(LR)}$	**0.0090**	51.19%	0.530	0.503	**0.0098**	52.17%	0.525	0.503	**0.0098**	51.89%	0.524	0.506
Stacking$_7^{(SVR)}$	0.0168	50.54%	0.523	0.503	0.0203	51.94%	0.515	0.513	0.0203	51.86%	0.518	0.512
Stacking$_7^{(kNN)}$	0.0165	52.00%	0.525	0.509	0.0179	52.66%	0.526	0.487	0.0100	51.91%	0.521	0.522
Averaging$_{11}$	0.0122	54.25%	**0.545**	0.542	0.0136	54.02%	**0.542**	**0.540**	0.0136	**54.03%**	0.501	**0.539**
Bagging$_{11}$	0.0124	**54.30%**	**0.555**	0.542	0.0134	**54.11%**	0.541	**0.540**	0.0093	53.11%	0.502	0.509
Stacking$_{11}^{(DTR)}$	0.0209	50.43%	0.510	0.502	0.0271	51.69%	0.510	0.512	0.0274	51.85%	0.519	0.512
Stacking$_{11}^{(LR)}$	**0.0087**	50.99%	0.529	0.492	**0.0095**	52.43%	0.522	0.507	0.0096	52.11%	0.514	0.501
Stacking$_{11}^{(SVR)}$	0.0165	50.66%	0.520	0.504	0.0202	51.85%	0.512	0.512	0.0202	51.82%	0.517	0.511
Stacking$_{11}^{(kNN)}$	0.0165	52.13%	0.526	0.505	0.0179	52.19%	0.523	0.489	0.0100	52.49%	**0.521**	0.520

Tables 5 and 6 present the performance of all ensemble models utilizing CNN-BiLSTM as base learner for $m = 4$ and $m = 9$, respectively. Firstly, it is worth noticing that stacking model using LR as meta-learner exhibited the best regression performance, regarding to all cryptocurrencies. Stacking$_7^{(LR)}$ and Stacking$_{11}^{(LR)}$ presented almost identical RMSE score Moreover, stacking$_7^{(LR)}$ presented slightly higher accuracy, AUC and F_1-score than stacking$_{11}^{(LR)}$, for ETH and XRP datasets, while for BTC dataset Stacking$_{11}^{(LR)}$ reported slightly better classification performance. This implies that the increment of component learners from 7 to 11 did not considerably improved and affected the regression and classification performance of the stacking ensemble algorithm. Stacking ensemble reported the worst (highest) RMSE score utilizing DTR, SVR and kNN as meta-learners. It is also worth mentioning that it exhibited the worst performance among all ensemble models and also worst than that of the single model CNN-BiLSTM. However, stacking ensemble reported the highest classification performance using kNN as meta-learner. Additionally, it presented slightly better classification performance using DTR or SVR than LR as meta-learners for ETH and XRP datasets, while for BTC dataset it presented better performance using LR as meta-learner as meta-learner.

Table 5. Performance of ensemble models using CNN and bi-directional LSTM (CNN-BiLSTM) as base learner for $m = 4$.

Model	BTC				ETH				XRP			
	RMSE	Acc	AUC	F_1	RMSE	Acc	AUC	F_1	RMSE	Acc	AUC	F_1
CNN-BiLSTM	0.0107	53.93%	0.528	0.522	0.0116	53.56%	0.525	0.521	0.0099	53.72%	0.519	0.484
Averaging$_7$	0.0104	54.15%	0.548	0.537	0.0114	53.87%	0.530	**0.536**	0.0096	54.73%	0.555	**0.532**
Bagging$_7$	0.0104	**54.62%**	0.549	0.544	0.0114	**53.88%**	0.532	0.529	0.0096	**55.42%**	0.557	0.520
Stacking$_7^{(DTR)}$	0.0176	50.33%	0.498	0.503	0.0224	51.20%	0.501	0.509	0.0155	51.66%	0.510	0.515
Stacking$_7^{(LR)}$	**0.0085**	51.67%	0.527	0.494	**0.0095**	52.81%	0.536	0.517	**0.0093**	51.04%	0.541	0.498
Stacking$_7^{(SVR)}$	**0.0085**	51.67%	0.510	0.494	0.0159	51.99%	0.513	0.517	0.0118	51.14%	0.511	0.510
Stacking$_7^{(kNN)}$	0.0112	52.18%	0.526	0.519	0.0126	54.15%	**0.544**	0.503	0.0101	52.56%	0.536	0.510
Averaging$_{11}$	0.0105	54.41%	0.546	**0.543**	0.0114	53.83%	0.531	**0.538**	0.0095	55.55%	0.525	**0.547**
Bagging$_{11}$	0.0104	**54.66%**	0.548	0.541	0.0113	54.07%	0.532	0.529	0.0096	**55.64%**	0.524	0.530
Stacking$_{11}^{(DTR)}$	0.0178	50.22%	0.494	0.502	0.0226	51.11%	0.510	0.508	0.0157	51.61%	0.518	0.514
Stacking$_{11}^{(LR)}$	**0.0085**	51.72%	0.527	0.499	**0.0095**	51.69%	0.533	0.492	**0.0092**	50.42%	**0.545**	0.496
Stacking$_{11}^{(SVR)}$	**0.0085**	51.72%	0.512	0.499	0.0161	52.36%	0.512	0.521	0.0116	51.82%	0.512	0.517
Stacking$_{11}^{(kNN)}$	0.0112	52.34%	0.521	0.504	0.0126	**54.14%**	0.542	0.492	0.0101	52.53%	0.529	0.509

Table 6. Performance of ensemble models using convolutional neural network with CNN-BiLSTM as base learner for $m = 9$.

Model	BTC				ETH				XRP			
	RMSE	Acc	AUC	F_1	RMSE	Acc	AUC	F_1	RMSE	Acc	AUC	F_1
CNN-BiLSTM	0.0146	53.47%	0.529	0.526	0.0154	53.87%	0.531	0.529	0.0101	50.36%	0.506	0.394
Averaging$_7$	0.0146	**53.58%**	**0.547**	**0.535**	0.0151	54.03%	**0.540**	**0.539**	0.0095	50.04%	0.537	0.428
Bagging$_7$	0.0143	53.20%	0.541	0.528	0.0155	**54.09%**	**0.540**	**0.539**	0.0095	51.11%	**0.540**	0.417
Stacking$_7^{(DTR)}$	0.0233	50.84%	0.521	0.507	0.0324	52.23%	0.522	0.517	0.0149	52.19%	0.519	**0.521**
Stacking$_7^{(LR)}$	**0.0091**	52.30%	0.518	0.516	**0.0105**	52.26%	0.525	0.512	**0.0093**	50.10%	0.522	0.488
Stacking$_7^{(SVR)}$	0.0109	52.30%	0.517	0.516	0.0235	52.10%	0.513	0.514	0.0112	51.28%	0.518	0.512
Stacking$_7^{(kNN)}$	0.0166	52.00%	0.522	0.508	0.0142	51.91%	0.529	0.475	0.0101	**52.50%**	0.523	0.510
Averaging$_{11}$	0.0144	**53.53%**	**0.548**	**0.535**	0.0151	54.02%	0.539	0.537	0.0093	50.22%	0.549	0.437
Bagging$_{11}$	0.0143	53.32%	0.547	0.530	0.0155	**54.20%**	**0.539**	**0.539**	0.0093	51.15%	**0.551**	0.470
Stacking$_{11}^{(DTR)}$	0.0235	50.76%	0.513	0.506	0.0305	52.41%	0.513	0.520	0.0152	51.72%	0.522	**0.516**
Stacking$_{11}^{(LR)}$	**0.0088**	50.48%	0.527	0.489	**0.0098**	52.67%	0.524	0.515	**0.0092**	50.68%	0.529	0.505
Stacking$_{11}^{(SVR)}$	0.0108	50.48%	0.513	0.489	0.0235	52.12%	0.516	0.514	0.0110	51.68%	0.517	**0.516**
Stacking$_{11}^{(kNN)}$	0.0166	52.00%	0.526	0.496	0.0142	51.95%	0.517	0.475	0.0100	**52.32%**	0.526	0.505

Regarding the other two ensemble strategies, averaging and bagging, they exhibited slightly better regression performance compared to the single CNN-BiLSTM model. Nevertheless, both averaging and bagging reported the highest accuracy, AUC and F_1-score, which implies that they presented the best classification performance among all other models with bagging exhibiting slightly better classification performance. Furthermore, it is also worth mentioning that both ensembles slightly improved their performance in term of RMSE score and Accuracy, when the number of component classifiers increased from 7 to 11.

In the follow-up, we provided a deeper insight classification performance of the forecasting models by presenting the confusion matrices of averaging$_{11}$, bagging$_{11}$, stacking$_7^{(LR)}$ and stacking$_{11}^{(kNN)}$ for $m = 4$, which exhibited the best overall performance. The use of the confusion matrix provides a compact and to the classification performance of each model, presenting complete information about mislabeled classes. Notice that each row of a confusion matrix represents the instances in an actual class while each column represents the instances in a predicted class. Additionally, both stacking ensembles utilizing DTR and SVM as meta-learners were excluded from the rest of our experimental analysis, since they presented the worst regression and classification performance, relative to all cryptocurrencies.

Tables 7–9 present the confusion matrices of the best identified ensemble models using CNN-LSTM as base learner, regarding BTC, ETH and XRP datasets, respectively. The confusion matrices for BTC and ETH revealed that stacking$_7^{(LR)}$ is biased, since most of the instances were misclassified as "Down", meaning that this model was unable to identify possible hidden patterns despite the fact that it exhibited the best regression performance. On the other hand, bagging$_{11}$ exhibited a balanced prediction distribution between "Down" or "Up" predictions, presenting its superiority over the rest forecasting models, followed by averaging$_{11}$ and stacking$_{11}^{(kNN)}$. Regarding XRP dataset, bagging$_{11}$ and stacking$_{11}^{(kNN)}$ presented the highest prediction accuracy and the best trade-off between true positive and true negative rate, meaning that these models may have identified some hidden patters.

Table 7. Confusion matrices of Averaging$_{11}$, Bagging$_{11}$, Stacking$_7^{(LR)}$ and Stacking$_{11}^{(kNN)}$ using CNN-LSTM as base model for Bitcoin (BTC) dataset.

Averaging$_{11}$			Bagging$_{11}$			Stacking$_7^{(LR)}$			Stacking$_{11}^{(kNN)}$		
	Down	Up		Down	Up		Down	Up		Down	Up
Down	1244	850	Down	1064	1030	Down	1537	557	Down	1296	798
Up	1270	1050	Up	922	1398	Up	1562	758	Up	1315	1005

Table 8. Confusion matrices of Averaging$_{11}$, Bagging$_{11}$, Stacking$_7^{(LR)}$ and Stacking$_{11}^{(kNN)}$ using CNN-LSTM as base model for Etherium (ETH) dataset.

Averaging$_{11}$			Bagging$_{11}$			Stacking$_7^{(LR)}$			Stacking$_{11}^{(kNN)}$		
	Down	Up		Down	Up		Down	Up		Down	Up
Down	1365	885	Down	1076	1174	Down	1666	584	Down	1353	897
Up	1208	955	Up	866	1297	Up	1593	570	Up	1203	960

Table 9. Confusion matrices of Averaging$_{11}$, Bagging$_{11}$, Stacking$_7^{(LR)}$ and Stacking$_{11}^{(kNN)}$ using CNN-LSTM as base model for Ripple (XRP) dataset.

Averaging$_{11}$			Bagging$_{11}$			Stacking$_7^{(LR)}$			Stacking$_{11}^{(kNN)}$		
	Down	Up		Down	Up		Down	Up		Down	Up
Down	1575	739	Down	1314	1000	Down	1072	1242	Down	1214	1100
Up	1176	923	Up	984	1115	Up	815	1284	Up	991	1108

Tables 10–12 present the confusion matrices of averaging$_{11}$, bagging$_{11}$, stacking$_7^{(LR)}$ and Stacking$_{11}^{(kNN)}$ using CNN-BiLSTM as base learner, regarding BTC, ETH and XRP datasets, respectively. The confusion matrices for BTC dataset demonstrated that both average$_{11}$ and bagging$_{11}$ presented the best performance while stacking$_7^{(LR)}$ was biased, since most of the instances were misclassified as "Down". Regarding ETH dataset, both average$_{11}$ and bagging$_{11}$ were considered biased since most "Up" instances were misclassified as "Down". In contrast, both stacking ensembles presented the best performance, with stacking$_{11}^{(kNN)}$ reporting slightly considerably better trade-off between sensitivity and specificity. Regarding XRP dataset, bagging$_{11}$ presented the highest prediction accuracy and the best trade-off between true positive and true negative rate, closely followed by stacking$_{11}^{(kNN)}$.

Table 10. Confusion matrices of Averaging$_{11}$, Bagging$_{11}$, Stacking$_7^{(LR)}$ and Stacking$_{11}^{(kNN)}$ using CNN-BiLSTM as base model for BTC dataset.

Averaging$_{11}$			Bagging$_{11}$			Stacking$_7^{(LR)}$			Stacking$_{11}^{(kNN)}$		
	Down	Up		Down	Up		Down	Up		Down	Up
Down	983	1111	Down	1019	1075	Down	1518	576	Down	1272	822
Up	864	1456	Up	915	1405	Up	1569	751	Up	1288	1032

Table 11. Confusion matrices of Averaging$_{11}$, Bagging$_{11}$, Stacking$_7^{(LR)}$ and Stacking$_{11}^{(kNN)}$ using CNN-BiLSTM as base model for ETH dataset.

Averaging$_{11}$			Bagging$_{11}$			Stacking$_7^{(LR)}$			Stacking$_{11}^{(kNN)}$		
	Down	Up		Down	Up		Down	Up		Down	Up
Down	1913	337	Down	1837	413	Down	1322	928	Down	1277	973
Up	1721	442	Up	1631	532	Up	1199	964	Up	1083	1080

Table 12. Confusion matrices of Averaging$_{11}$, Bagging$_{11}$, Stacking$_7^{(LR)}$ and Stacking$_{11}^{(kNN)}$ using CNN-BiLSTM as base model for XRP dataset.

Averaging$_{11}$			Bagging$_{11}$			Stacking$_7^{(LR)}$			Stacking$_{11}^{(kNN)}$		
	Down	Up		Down	Up		Down	Up		Down	Up
Down	1518	796	Down	1440	874	Down	1124	1190	Down	1221	1093
Up	1146	953	Up	1062	1037	Up	831	1268	Up	998	1101

In the rest of this section, we evaluate the reliability of the best reported ensemble models by examining if they have properly fitted the time series. In other words, we examine if the models' residuals defined by

$$\hat{e}_t = y_t - \hat{y}_t$$

are identically distributed and asymptotically independent. It is worth noticing the residuals are dedicated to evaluate whether the model has properly fitted the time series.

For this purpose, we utilize the AutoCorrelation Function (ACF) plot [47] which is obtained from the linear correlation of each residual $\hat{e}_t$ to the others in different lags, $\hat{e}_{t-1}, \hat{e}_{t-2}, \ldots$ and illustrates the intensity of the temporal autocorrelation. Notice that in case the forecasting model violates the assumption of no autocorrelation in the errors implies that its predictions may be inefficient since there is some additional information left over which should be accounted by the model.

Figures 4–6 present the ACF plots for BTC, ETH and XRP datasets, respectively. Notice that the confident limits (blue dashed line) are constructed assuming that the residuals follow a Gaussian probability distribution. It is worth noticing that averaging$_{11}$ and bagging$_{11}$ ensemble models violate the assumption of no autocorrelation in the errors which suggests that their forecasts may be inefficient, regarding BTC and ETH datasets. More specifically, the significant spikes at lags 1 and 2 imply that there exists some additional information left over which should be accounted by the models. Regarding XRP dataset, the ACF plot of average$_{11}$ presents that the residuals have no autocorrelation; while the ACF plot of bagging$_{11}$ presents that there is a spike at lag 1, which violates the assumption of no autocorrelation in the residuals. Both ACF plots of stacking ensemble are within 95% percent confidence interval for all lags, regarding BTC and XRP datasets, which verifies that the residuals have no autocorrelation. Regarding the ETH dataset, the ACF plot of stacking$_7^{(LR)}$ reported a small spike at lag 1, which reveals that there is some autocorrelation of the residuals but not particularly large; while the ACF plot of stacking$_{11}^{(kNN)}$ reveals that there exist small spikes at lags 1 and 2, implying that there is some autocorrelation.

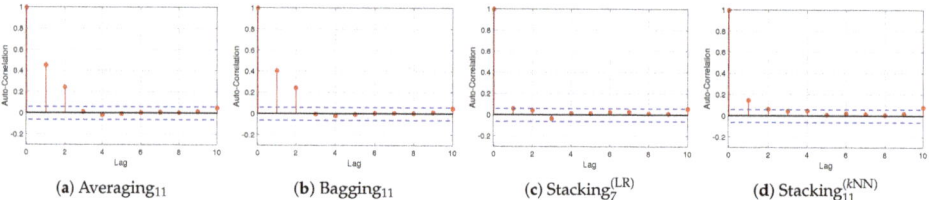

Figure 4. Autocorrelation of residuals for BTC dataset of ensemble models using CNN-LSTM as base learner.

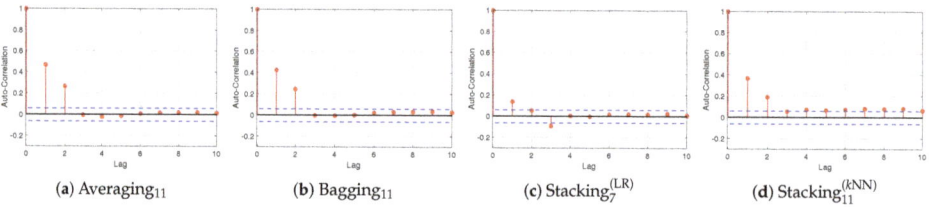

Figure 5. Autocorrelation of residuals for ETH dataset of ensemble models using CNN-LSTM as base learner.

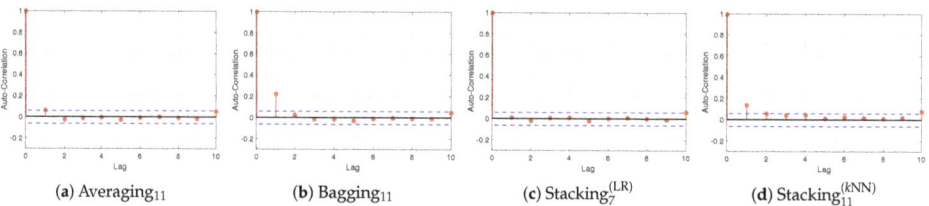

Figure 6. Auto-correlation of residuals for XRP dataset of ensemble models using CNN-LSTM as base learner.

Figures 7–9 present the ACF plots of averaging$_{11}$, bagging$_{11}$, stacking$_7^{(LR)}$ and stacking$_{11}^{(kNN)}$ ensembles utilizing CNN-BiLSTM as base learner for BTC, ETH and XRP datasets, respectively. Both averaging$_{11}$ and bagging$_{11}$ ensemble models violate the assumption of no autocorrelation in the errors, relative to all cryptocurrencies, implying that these models are not properly fitted the time-series. In more detail, the significant spikes at lags 1 and 2 suggest that the residuals are not identically distributed and asymptotically independent, for all datasets. The ACF plot of stacking$_7^{(LR)}$ ensemble for BTC dataset verify that the residuals have no autocorrelation since are within 95% percent confidence interval for all lags. In contrast, for ETH and XRP datasets the spikes at lags 1 and 2 illustrate that there is some autocorrelation of the residuals. Regarding the ACF plots of stacking$_{11}^{(kNN)}$ present that there exists some autocorrelation in the residuals but not particularly large for BTC and XRP datasets; while for ETH dataset, the significant spikes at lags 1 and 2 suggest that the model's prediction may be inefficient.

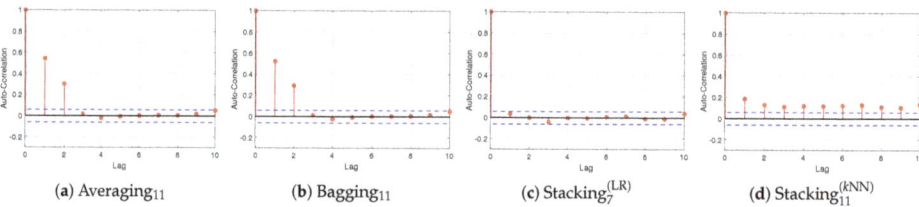

Figure 7. Autocorrelation of residuals for BTC dataset of ensemble models using CNN-BiLSTM as base learner.

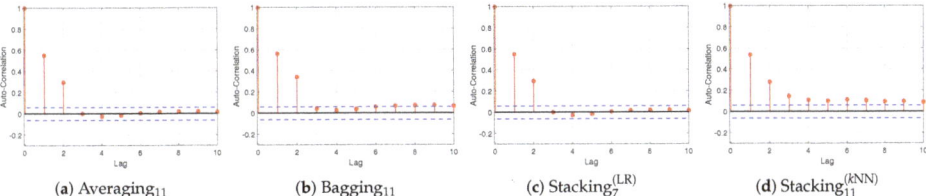

Figure 8. Autocorrelation of residuals for ETH dataset of ensemble models using CNN-BiLSTM as base learner.

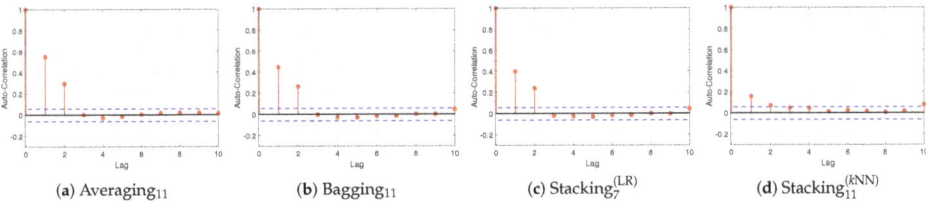

Figure 9. Autocorrelation of residuals for XRP dataset of ensemble models using CNN-BiLSTM as base learner.

6. Discussion

In this section, we perform a discussion regarding the proposed ensemble models, the experimental results and the main finding of this work.

6.1. Discussion of Proposed Methodology

Cryptocurrency prediction is considered a very challenging forecasting problem, since the historical prices follow a random walk process, characterized by large variations in the volatility, although a few hidden patterns may probably exist [48,49]. Therefore, the investigation and the development of a powerful forecasting model for assisting decision making and investment policies is considered essential. In this work, we incorporated advanced deep learning models as base learners into three of the most popular and widely used ensembles methods, namely averaging, bagging and stacking for forecasting cryptocurrency hourly prices.

The motivation behind our approach is to exploit the advantages of ensemble learning and advanced deep learning techniques. More specifically, we aim to exploit the effectiveness of ensemble learning for reducing the bias or variance of error by exploiting multiple learners and the ability of deep learning models to learn the internal representation of the cryptocurrency data. It is worth mentioning that since the component deep learning learners are initialized with different weight states, this leads to the development of deep learning models each of which focuses on different identified patterns. Therefore, the combination of these learners via an ensemble learning strategy may lead to stable and robust prediction model.

In general, deep learning neural networks are powerful prediction models in terms of accuracy, but are usually unstable in sense that variations in their training set or in their weight initialization may significantly affect their performance. Bagging strategy constitutes an effective way of building efficient and stable prediction models, utilizing unstable and diverse base learners [50,51], aiming to reduce variance and avoid overfitting. In other words, bagging stabilizes the unstable deep learning base learners and exploits their prediction accuracy focusing on building an accurate and robust final prediction model. However, the main problem of this approach is that since bagging averages the predictions of all models, redundant and non-informative models may add too much noise on the final prediction result and therefore, possible identified patterns, by some informative and valuable models, may disappear.

On the other hand, stacking ensemble learning utilizes a meta-learner in order to learn the prediction behavior of the base learners, with respect to the final target output. Therefore, it is able to identify the redundant and informative base models and "weight them" in a nonlinear and more intelligent way in order to filter out useless and non-informative base models. As a result, the selection of the meta-learner is of high significance for the effectiveness and efficiency of this ensemble strategy.

6.2. Discussion of Results

All compared ensemble models were evaluated considering both regression and classification problems, namely for the prediction of the cryptocurrency price on the following hour (regression) and also for the prediction if the price will increase or decrease on the following hour (classification). Our experiments revealed that the incorporation of deep learning models into ensemble learning framework improved the prediction accuracy in most cases, compared to a single deep learning model.

Bagging exhibited the best overall score in terms of classification accuracy, closely followed by averaging and stacking$^{(kNN)}$; while stacking$^{(LR)}$ the best regression performance. The confusion matrices revealed that stacking$^{(LR)}$ as base learner was actually biased, since most of the instances were wrongly classified as "Down" while bagging and stacking$^{(kNN)}$ exhibited a balanced prediction distribution between "Down" or "Up" predictions. It is worth noticing that since bagging can be interpreted as a perturbation technique aiming at improving the robustness especially against outliers and highly volatile prices [37]. The numerical experiments demonstrated that averaging ensemble models trained on perturbed training dataset is a means to favor invariance to these perturbations and better capture the directional movements of the presented random walk processes. However, the ACF plots revealed that bagging ensemble models violate the assumption of no autocorrelation in the residuals, which implies that their predictions may be inefficient. In contrast, the ACF plots of stacking$^{(kNN)}$ revealed that the residuals have no or small (inconsiderable) autocorrelation. This is probably due to fact that the use of a meta-learner, which is trained on the errors of the base learners, is able to reduce the autocorrelation in the residuals and provide more reliable forecasts. Finally, it is worth mentioning that the increment of component learners had little or no effect to the regression performance of the ensemble algorithms, in most cases

Summarizing, stacking utilizing advanced deep learning base learner and kNN as meta-learner may considered to be the best forecasting model for the problem of cryptocurrency price prediction and movement, based on our experimental analysis. Nevertheless, further research has to be performed in order to improve the prediction performance of our prediction framework by creating even more innovative and sophisticated algorithmic models. Moreover, additional experiments with respect to the trading-investment profit returns based on such prediction frameworks have to be also performed.

7. Conclusions

In this work, we explored the adoption of ensemble learning strategies with advanced deep learning models for forecasting cryptocurrency price and movement, which constitutes the main contribution of this research. The proposed ensemble models utilize state-of-the-art deep learning models as component learners, which are based on combinations of LSTM, BiLSTM and convolutional layers. An extensive and detailed experimental analysis was performed considering both classification and regression performance evaluation of averaging, bagging, and stacking ensemble strategies. Furthermore, the reliability and the efficiency of the predictions of each ensemble model was studied by examining for autocorrelation of the residuals.

Our numerical experiments revealed that ensemble learning and deep learning may efficiently be adapted to develop strong, stable, and reliable forecasting models. It is worth mentioning that due to the sensitivity of various hyper-parameters of the proposed ensemble models and their high complexity, it is possible that their prediction ability could be further improved by performing additional optimized configuration and mostly feature engineering. Nevertheless, in many real-world applications, the selection of the base learner as well as the specification of their number in an ensemble strategy constitute a significant choice in terms of prediction accuracy, reliability, and computation time/cost. Actually, this fact acts as a limitation of our approach. The incorporation of deep learning models (which are by nature computational inefficient) in an ensemble learning approach, would lead the total training and prediction computation time to be considerably increased. Clearly, such an ensemble model would be inefficient on real-time and dynamic applications tasks with high-frequency inputs/outputs, compared to a single model. However, on low-frequency applications when the objective is the accuracy and reliability, such a model could significantly shine.

Our future work is concentrated on the development of an accurate and reliable decision support system for cryptocurrency forecasting enhanced with new performance metrics based on profits and returns. Additionally, an interesting idea which is worth investigating in the future is that in certain times of global instability, we experience a significant number of outliers in the prices of all cryptocurrencies. To address this problem an intelligent system might be developed based on an anomaly detection framework, utilizing unsupervised algorithms in order to "catch" outliers or other rare signals which could indicate cryptocurrency instability.

Author Contributions: Supervision, P.P.; Validation, E.P.; Writing—review & editing, I.E.L. and S.S. All authors have read and agreed to the published version of the manuscript.

Funding: This research received no external funding.

Conflicts of Interest: The authors declare no conflict of interest.

List of Acronyms and Abbreviations

Acronym	Description
BiLSTM	Bi-directional Long Short-Term Memory
BRNN	Bi-directional Recurrent Neural Network
BTC	Bitcoin
CNN	Convolutional Neural Network
DTR	Decision Tree Regression
ETH	Ethereum
F_1	F_1-score
kNN	k-Nearest Neighbor
LR	Linear Regression
LSTM	Long Short-Term Memory
MLP	Multi-Layer Perceptron
SVR	Support Vector Regression
RMSE	Root Mean Square Error
XRP	Ripple

References

1. Nakamoto, S. Bitcoin: A Peer-to-Peer Electronic Cash System. 2008. Available online: https://git.dhimmel.com/bitcoin-whitepaper/ (accessed on 20 February 2020).
2. De Luca, G.; Loperfido, N. A Skew-in-Mean GARCH Model for Financial Returns. In *Skew-Elliptical Distributions and Their Applications: A Journey Beyond Normality*; Corazza, M., Pizzi, C., Eds.; CRC/Chapman & Hall: Boca Raton, FL, USA, 2004; pp. 205–202.
3. De Luca, G.; Loperfido, N. Modelling multivariate skewness in financial returns: A SGARCH approach. *Eur. J. Financ.* **2015**, *21*, 1113–1131. [CrossRef]
4. Weigend, A.S. *Time Series Prediction: Forecasting the Future and Understanding the Past*; Routledge: Abingdon, UK, 2018.
5. Azoff, E.M. *Neural Network Time Series Forecasting of Financial Markets*; John Wiley & Sons, Inc.: Hoboken, NJ, USA, 1994.
6. Oancea, B.; Ciucu, Ş.C. Time series forecasting using neural networks. *arXiv* **2014**, arXiv:1401.1333.

7. Pintelas, E.; Livieris, I.E.; Stavroyiannis, S.; Kotsilieris, T.; Pintelas, P. *Fundamental Research Questions and Proposals on Predicting Cryptocurrency Prices Using DNNs*; Technical Report TR20-01; University of Patras: Patras, Greece, 2020. Available online: https://nemertes.lis.upatras.gr/jspui/bitstream/10889/13296/1/TR01-20.pdf (accessed on 20 February 2020).
8. Wen, Y.; Yuan, B. Use CNN-LSTM network to analyze secondary market data. In Proceedings of the 2nd International Conference on Innovation in Artificial Intelligence, Shanghai, China, 9–12 March 2018; pp. 54–58.
9. Liu, S.; Zhang, C.; Ma, J. CNN-LSTM neural network model for quantitative strategy analysis in stock markets. In *International Conference on Neural Information Processing*; Springer: Berlin/Heidelberg, Germany, 2017; pp. 198–206.
10. Livieris, I.E.; Pintelas, E.; Pintelas, P. A CNN-LSTM model for gold price time-series forecasting. *Neural Comput. Appl.* Available online: https://link.springer.com/article/10.1007/s00521-020-04867-x (accessed on 20 February 2020). [CrossRef]
11. Pintelas, E.; Livieris, I.E.; Stavroyiannis, S.; Kotsilieris, T.; Pintelas, P. Investigating the problem of cryptocurrency price prediction—A deep learning approach. In Proceedings of the 16th International Conference on Artificial Intelligence Applications and Innovations, Neos Marmaras, Greece, 5–7 June 2020.
12. Yiying, W.; Yeze, Z. Cryptocurrency Price Analysis with Artificial Intelligence. In Proceedings of the 2019 5th International Conference on Information Management (ICIM), Cambridge, UK, 24–27 March 2019; pp. 97–101.
13. Nakano, M.; Takahashi, A.; Takahashi, S. Bitcoin technical trading with artificial neural network. *Phys. A Stat. Mech. Its Appl.* **2018**, *510*, 587–609. [CrossRef]
14. McNally, S.; Roche, J.; Caton, S. Predicting the price of Bitcoin using Machine Learning. In Proceedings of the 2018 26th Euromicro International Conference on Parallel, Distributed and Network-based Processing (PDP), Cambridge, UK, 21–23 March 2018; pp. 339–343.
15. Shintate, T.; Pichl, L. Trend prediction classification for high frequency bitcoin time series with deep learning. *J. Risk Financ. Manag.* **2019**, *12*, 17. [CrossRef]
16. Miura, R.; Pichl, L.; Kaizoji, T. Artificial Neural Networks for Realized Volatility Prediction in Cryptocurrency Time Series. In *International Symposium on Neural Networks*; Springer: Berlin/Heidelberg, Germany, 2019; pp. 165–172.
17. Ji, S.; Kim, J.; Im, H. A Comparative Study of Bitcoin Price Prediction Using Deep Learning. *Mathematics* **2019**, *7*, 898. [CrossRef]
18. Kumar, D.; Rath, S. Predicting the Trends of Price for Ethereum Using Deep Learning Techniques. In *Artificial Intelligence and Evolutionary Computations in Engineering Systems*; Springer: Berlin/Heidelberg, Germany, 2020; pp. 103–114.
19. Hochreiter, S.; Schmidhuber, J. Long short-term memory. *Neural Comput.* **1997**, *9*, 1735–1780. [CrossRef] [PubMed]
20. Yu, Y.; Si, X.; Hu, C.; Zhang, J. A review of recurrent neural networks: LSTM cells and network architectures. *Neural Comput.* **2019**, *31*, 1235–1270. [CrossRef] [PubMed]
21. Zhang, K.; Chao, W.L.; Sha, F.; Grauman, K. Video summarization with long short-term memory. In *European Conference on Computer Vision*; Springer: Berlin/Heidelberg, Germany, 2016; pp. 766–782.
22. Nowak, J.; Taspinar, A.; Scherer, R. LSTM recurrent neural networks for short text and sentiment classification. In *International Conference on Artificial Intelligence and Soft Computing*; Springer: Berlin/Heidelberg, Germany, 2017; pp. 553–562.
23. Rahman, L.; Mohammed, N.; Al Azad, A.K. A new LSTM model by introducing biological cell state. In Proceedings of the 2016 3rd International Conference on Electrical Engineering and Information Communication Technology (ICEEICT), Dhaka, Bangladesh, 22–24 September 2016; pp. 1–6.
24. Schuster, M.; Paliwal, K.K. Bidirectional recurrent neural networks. *IEEE Trans. Signal Process.* **1997**, *45*, 2673–2681. [CrossRef]
25. Graves, A.; Mohamed, A.R.; Hinton, G. Speech recognition with deep recurrent neural networks. In Proceedings of the 2013 IEEE International Conference on Acoustics, Speech and Signal Processing, Vancouver, BC, Canada, 26–31 May 2013; pp. 6645–6649.
26. Lu, L.; Wang, X.; Carneiro, G.; Yang, L. *Deep Learning and Convolutional Neural Networks for Medical Imaging and Clinical Informatics*; Springer: Berlin/Heidelberg, Germany, 2019.

27. Rawat, W.; Wang, Z. Deep convolutional neural networks for image classification: A comprehensive review. *Neural Comput.* **2017**, *29*, 2352–2449. [CrossRef] [PubMed]
28. Michelucci, U. *Advanced Applied Deep Learning: Convolutional Neural Networks and Object Detection*; Springer: Berlin/Heidelberg, Germany, 2019.
29. Ioffe, S.; Szegedy, C. Batch normalization: Accelerating deep network training by reducing internal covariate shift. *arXiv* **2015**, arXiv:1502.03167.
30. Srivastava, N.; Hinton, G.; Krizhevsky, A.; Sutskever, I.; Salakhutdinov, R. Dropout: A simple way to prevent neural networks from overfitting. *J. Mach. Learn. Res.* **2014**, *15*, 1929–1958.
31. Rokach, L. *Ensemble Learning: Pattern Classification Using Ensemble Methods*; World Scientific Publishing Co Pte Ltd.: Singapore, 2019.
32. Lior, R. *Ensemble Learning: Pattern Classification Using Ensemble Methods*; World Scientific: Singapore, 2019; Volume 85.
33. Zhou, Z.H. *Ensemble Methods: Foundations and Algorithms*; Chapman & Hall/CRC: Boca Raton, FL, USA, 2012.
34. Bian, S.; Wang, W. On diversity and accuracy of homogeneous and heterogeneous ensembles. *Int. J. Hybrid Intell. Syst.* **2007**, *4*, 103–128. [CrossRef]
35. Polikar, R. Ensemble learning. In *Ensemble Machine Learning*; Springer: Berlin/Heidelberg, Germany, 2012; pp. 1–34.
36. Christiansen, B. Ensemble averaging and the curse of dimensionality. *J. Clim.* **2018**, *31*, 1587–1596. [CrossRef]
37. Grandvalet, Y. Bagging equalizes influence. *Mach. Learn.* **2004**, *55*, 251–270. [CrossRef]
38. Livieris, I.E.; Iliadis, L.; Pintelas, P. On ensemble techniques of weight-constrained neural networks. *Evol. Syst.* **2020**, 1–13. [CrossRef]
39. Wolpert, D.H. Stacked generalization. *Neural Netw.* **1992**, *5*, 241–259. [CrossRef]
40. Gulli, A.; Pal, S. *Deep Learning with Keras*; Packt Publishing Ltd.: Birmingham, UK, 2017.
41. Kingma, D.P.; Ba, J. Adam: A method for stochastic optimization. In Proceedings of the 2015 International Conference on Learning Representations, San Diego, CA, USA, 7–9 May 2015.
42. Wu, X.; Kumar, V. *The Top Ten Algorithms in Data Mining*; CRC Press: Boca Raton, FL, USA, 2009.
43. Deng, N.; Tian, Y.; Zhang, C. *Support Vector Machines: Optimization Based Theory, Algorithms, and Extensions*; Chapman and Hall/CRC: Boca Raton, FL, USA, 2012.
44. Kutner, M.H.; Nachtsheim, C.J.; Neter, J.; Li, W. *Applied Linear Statistical Models*; McGraw-Hill Irwin: New York, NY, USA, 2005; Volume 5.
45. Aha, D.W. *Lazy learning*; Springer Science & Business Media: Berlin/Heidelberg, Germany, 2013.
46. Breiman, L.; Friedman, J.; Olshen, R. *Classification and Regression Trees*; Routledge: Abingdon, UK, 2017.
47. Brockwell, P.J.; Davis, R.A. *Introduction to Time Series and Forecasting*; Springer: Berlin/Heidelberg, Germany, 2016.
48. Stavroyiannis, S. Can Bitcoin diversify significantly a portfolio? *Int. J. Econ. Bus. Res.* **2019**, *18*, 399–411. [CrossRef]
49. Stavroyiannis, S. Value-at-Risk and Expected Shortfall for the major digital currencies. *arXiv* **2017**, arXiv:1708.09343.
50. Elisseeff, A.; Evgeniou, T.; Pontil, M. Stability of randomized learning algorithms. *J. Mach. Learn. Res.* **2005**, *6*, 55–79.
51. Kotsiantis, S.B. Bagging and boosting variants for handling classifications problems: A survey. *Knowl. Eng. Rev.* **2014**, *29*, 78–100. [CrossRef]

© 2020 by the authors. Licensee MDPI, Basel, Switzerland. This article is an open access article distributed under the terms and conditions of the Creative Commons Attribution (CC BY) license (http://creativecommons.org/licenses/by/4.0/).

MDPI
St. Alban-Anlage 66
4052 Basel
Switzerland
Tel. +41 61 683 77 34
Fax +41 61 302 89 18
www.mdpi.com

Algorithms Editorial Office
E-mail: algorithms@mdpi.com
www.mdpi.com/journal/algorithms

www.ingramcontent.com/pod-product-compliance
Lightning Source LLC
LaVergne TN
LVHW070651100526
838202LV00013B/938